T0134566

Secure Networked Inference with Unreliable Data Sources

Aditya Vempaty · Bhavya Kailkhura
Pramod K. Varshney

Secure Networked Inference with Unreliable Data Sources

Aditya Vempaty
IBM Research—Thomas J. Watson Research
Yorktown Heights, NY, USA

Bhavya Kailkhura
Department of Computing Applications and Research
Lawrence Livermore National Laboratory
Livermore, CA, USA

Pramod K. Varshney
Department of Electrical Engineering and Computer Science
Syracuse University
Syracuse, NY, USA

ISBN 978-981-13-4765-8 ISBN 978-981-13-2312-6 (eBook)
https://doi.org/10.1007/978-981-13-2312-6

Softcover re-print of the Hardcover 1st edition 2018

This Springer imprint is published by the registered company Springer Nature Singapore Pte Ltd.
The registered company address is: 152 Beach Road, #21-01/04 Gateway East, Singapore 189721, Singapore

To our families

My parents Anil and Radha
My wife Swetha

Aditya Vempaty

My parents Umesh and Anu
My brother Lakshya

Bhavya Kailkhura

My wife Anju

Pramod K. Varshney

Preface

With an explosion in the number of connected devices and the emergence of big and dirty data era, new distributed learning solutions are needed to tackle the problem of inference with corrupted data. The aim of this book is to present theory and algorithms for secure networked inference in the presence of unreliable data sources. More specifically, we present fundamental limits of networked inference in the presence of Byzantine data (malicious data sources) and discuss robust mitigation strategies to ensure reliable performance for several practical network architectures. In particular, the inference (or learning) process can be detection, estimation, or classification, and the network architecture of the system can be parallel, hierarchical, or fully decentralized (peer to peer).

This book is a result of our active research over the past decade, where we have been working on the generalization and modernization of the classical "Byzantine generals problem" in the context of distributed inference to different network topologies. Over the last three decades, the research community has extensively studied the impact of imperfect transmission channels or sensor faults on distributed inference systems. However, unreliable (Byzantine) data models considered in this book are philosophically different from the imperfect channels or faulty sensor cases. Byzantines are intentional and intelligent and therefore can strategically optimize their actions over the data corruption parameters. Thus, in contrast to channel-aware inference, both the network and Byzantines can optimize their utility by choosing their actions based on the knowledge of their opponent's behavior, leading to a game between the network and Byzantines. Study of these practically motivated scenarios in the presence of Byzantines is of utmost importance and is missing from the channel-aware and fault-tolerant inference literature.

This book provides a thorough introduction to various aspects of recent advances in secure distributed inference in the presence of Byzantines for various network topologies. Researchers, engineers, and graduate students who are interested in statistical inference, wireless networks, wireless sensor networks, network security, data/information fusion, distributed surveillance, or distributed signal processing will benefit from this book. After reading this book, they are expected to understand the principles behind distributed inference especially in the presence of Byzantines, learn advanced optimization techniques to solve inference problems in the presence of practical constraints such as imperfect data, and get insights on inference in ad hoc networks through practical examples. This book will lay a foundation for readers to solve real-world networked inference problems. It will be valuable to students, researchers, and practicing engineers at universities, government research laboratories, and companies who are interested in developing robust and secure inference solutions for real-world ad hoc inference networks. The covered theory will be useful for a wide variety of applications such as military surveillance, distributed spectrum sensing (DSS), source localization and tracking, traffic and environment monitoring, IoT, and crowdsourcing.

Readers should have some background knowledge of probability theory, statistical signal processing, and optimization. Familiarity with distributed detection and data fusion (e.g., see *P. K. Varshney, Distributed Detection and Data Fusion. New York: Springer-Verlag, 1997.*) will be helpful for some topics, but is not required to understand the majority of the content.

Yorktown Heights, USA — Aditya Vempaty
Livermore, USA — Bhavya Kailkhura
Syracuse, USA — Pramod K. Varshney

Acknowledgements

This book owes a considerable debt of gratitude to those who have influenced our work and our thinking over the years.

We are grateful to Hao Chen who initiated this research direction in Sensor Fusion Laboratory at Syracuse University. The material presented in Sect. 6.2 was originally conceived by Lav Varshney. We would like to thank Keshav Agrawal, Priyank Anand, Swastik Brahma, Berkan Dulek, Mukul Gagrani, Yunghsiang S. Han, Wael Hashlamoun, Satish Iyengar, Qunwei Li, Sid Nadendla, Onur Ozdemir, Ankit Rawat, Priyadip Ray, and Pranay Sharma for actively contributing to the research results presented in this book. Kush Varshney and Lav Varshney have provided critical comments and suggestions during this entire research.

We are also extremely blessed to have supportive families who inspired and encouraged us to complete this book.

Aditya Vempaty would like to thank his family—parents Anil Kumar Vempaty and Radha Vempaty, and wife Swetha Stotra Bhashyam—for their unconditional love, continued support, and extreme patience. Their presence has been a great mental support and is greatly appreciated.

Bhavya Kailkhura would like to thank his parents, grandparents, brother, and the rest of the family for their unconditional love, support, and sacrifices throughout his life.

Pramod K. Varshney is extremely blessed to have an extremely upbeat wife who inspired him throughout his career. He dedicates the research that led to this book to his life partner for 40 years, Anju Varshney. She faced severe adversities but tackled them with a positive attitude. Her motto has been *Life Happens Keep*

Smiling. His children Lav, Kush, Sonia, and Nina have provided constant encouragement. His grandchildren Arya, Ribhu, and Kaushali are simply bundles of joy and have made his life much happier and cheerful.

July 2018

Aditya Vempaty
Bhavya Kailkhura
Pramod K. Varshney

Contents

Chapter 1
Introduction

We are living in an increasingly networked world with sensing networks that are often comprised of numerous sensing devices (or agents) of varying sizes, communicating with each other via different topologies. Inferring about an unknown phenomenon based on observations of multiple agents has become a part of our daily lives. Almost every decision we make is made after collecting evidence from multiple distributed sources (agents). The problem of distributed inference is now a classical problem in statistics and engineering [3, 4] and is fairly complex. To make the problem even more complicated, the agents in the network can be unreliable due to a variety of reasons such as noise, faults, and malicious attacks, thus providing unreliable data. Distributed inference was originally motivated by its applications in military surveillance but is now being employed in a wide variety of applications such as search and rescue using drones, distributed spectrum sensing (DSS) using cognitive radio networks (CRNs), cyber-physical systems including Internet of things (IoT), and traffic and environment monitoring. Although the area of statistical inference has been an active area of research in the past, distributed learning and inference in a networked setup with potentially unreliable components and information sources has gained attention only recently. The emergence of big and dirty data era demands new distributed learning and inference solutions to tackle the problem of inference with unreliable data, the topic addressed in this book.

1.1 Distributed Inference Networks

Distributed inference networks (DINs) consist of a group of networked agents which acquire observations regarding a phenomenon of interest (POI), and collaborate with the other agents in the network by sharing their inferences via different topologies to make a global inference. Specifically, the inference (or learning) process can be that of detection, estimation, or classification, and the topology of the system

A. Vempaty et al., *Secure Networked Inference with Unreliable Data Sources*,
https://doi.org/10.1007/978-981-13-2312-6_1

can be parallel, hierarchical, or fully decentralized (peer to peer). The motivation behind the frequent use of a large number of inexpensive and less reliable agents is to provide a balance between cost and functionality. The inference performance of such systems strongly depends on the reliability of agents in the network. While there are several practical challenges one faces while inferring based on data provided by multiple agents in DINs, this book focuses on the class of challenges associated with the presence of unreliable data (observations/information). Causes of unreliable data can be divided into three types: (1) channel noise, (2) faulty sensors, and (3) malicious attackers. The major focus of this book is on the unreliability introduced by cause (3) where an adversary intentionally injects corrupted data to degrade the inference performance of the system.[1] Note that the unreliable data model may seem similar to the scenario where local decisions are transmitted over a binary symmetric channel (BSC) with a certain crossover probability. However, as discussed later there are fundamental differences. Over the last three decades, the research community has extensively studied the impact of transmission channels or faults on the distributed detection system and related problems due to its importance in several applications. However, unreliability caused by Byzantines considered in this book is philosophically different from the BSC and the faulty sensor cases. Byzantines are both intentional and intelligent, and therefore, they can optimize over the data corruption parameters. In contrast to channel-aware detection, both the fusion center (FC) of a DIN and the Byzantines can optimize their utility by choosing their actions based on the knowledge of the behavior of their opponents. Study of these practically motivated scenarios in the presence of Byzantines is of utmost importance and is missing from the channel-aware and fault-tolerant inference literature. This book advances the distributed inference literature by discussing fundamental limits of distributed inference with Byzantine data and provides optimal countermeasures (using the insights provided by these fundamental limits) from a network designer's perspective. This research problem is motivated by the Byzantine generals problem [1] as discussed next.

1.2 Motivation: Byzantine Generals' Problem

In 1982, Lamport et al. [1] presented the so-called Byzantine generals problem as follows: "…a group of generals of the Byzantine army camped with their troops around an enemy city. Communicating only by messengers, the generals must agree upon a common battle plan. However, one or more of them may be traitors who will try to mislead the others. The problem is to find an algorithm to ensure that the loyal generals will reach agreement." This problem is similar in principle to the problem

[1]The adversary node is called a Byzantine, and its action is called Byzantine action of producing corrupted data. More details regarding Byzantines are provided later in this chapter.

considered in this book. The authors gave a sharp characterization of the power of the Byzantine generals. It was shown that if the fraction of Byzantine generals is less than 1/3, there is a way for the loyal generals to reach a consensus agreement, regardless of what the Byzantine generals do. If the fraction is above 1/3, consensus can no longer be guaranteed.

It is not difficult to relate the Byzantine generals problem to the problem of secure distributed inference, where Byzantine generals play the role of internal adversaries within the DIN providing unreliable data. There are many diverse behaviors that a Byzantine entity may engage in; e.g., an agent may lie about connectivity, flood the network with false traffic, attempt to subjugate control information, falsely describe opinions of another node, or capture a strategic subset of devices and collude. This book examines the Byzantine generals problem in the context of distributed inference, where data collected by agents at remote locations is processed for specific inference purposes. The assumption is that the data sources are potentially tampered by some internal adversaries who have the knowledge about the algorithm used for inference and corrupt/falsify agent data. In this book, we will refer to the problem considered as the problem of distributed inference with Byzantine data.

Next, we describe two viewpoints adopted in the book for analyzing DINs in the presence of Byzantines.

1.3 Distributed Inference in the Presence of Byzantines: Two Viewpoints

Analysis of networked inference systems is much more challenging compared to classical inference systems that ignore the data generation pipeline where the data could be tampered with. In a distributed inference network, both the adversary and the network designer have several degrees of freedom which makes the problem complicated. In this book, we study the problem of distributed inference with Byzantine data from both attacker's and network designer's perspectives. Note that finding the optimal attacking strategies is the first step toward designing a robust distributed inference system in a holistic manner.

1.3.1 Analysis from the Attacker's Perspective

From the Byzantine's perspective, one problem of interest is to characterize the degree to which they can affect the inference performance of DINs. We are interested in the minimum fraction of Byzantine agents that make the inference by the DIN no better than one that is based only on prior information without using any sensed data. This is referred to as blinding the network. We refer to the fraction of nodes that are

Byzantines in the network[2] as the attack power of the Byzantines and minimum power required to blind the network as "critical power." Some important questions from a Byzantine's perspective to be answered are:

- How does network topology affect the critical power? In other words, which network architectures are more susceptible to Byzantine unreliabilities?
- Can we analytically characterize the critical power in different networks?
- If the power of the adversary is less than the critical value, what should be the optimal attacking strategy of the adversary?
- Finally, in a heterogeneous inference network, which resources/nodes should a cost-constrained adversary attack to maximize its impact?

While these questions are difficult to answer in general, some insights can be obtained by utilizing tools from hypothesis testing, estimation theory, game theory, and information theory.

1.3.2 Analysis from the Network Designer's Perspective

The previous discussion considers the issue of inference based on unreliable data from the Byzantine's perspective. However, one needs to look at the possible countermeasures from the network designer's perspective to protect the network from these Byzantines. We follow the methodology suggested by Claude Shannon in his unpublished manuscript of 1956 titled "Reliable Machines from Unreliable Components" which considers the problem of designing reliable machines from unreliable components [2]. He suggests that there are typically three methods to improve system reliability: (1) improve individual system components, (2) use of error-correcting codes, and (3) complete system redesign. We employ all three approaches as follows. System components are improved either by identifying the local malicious nodes and using them for further inference and/or improving the performance of the global detector by implementing the optimal fusion rule. Although the cause of unreliable information is different for faulty sensors and Byzantines, the effect is the same; namely, errors are introduced in the data. Therefore, coding theory ideas are also used to correct these errors and improve inference performance. And finally, as seen later in the book, complete redesign corresponds to a total change in the inference architecture. The problem of optimal network structure design which is robust to Byzantines is solved in a distributed inference context.

[2] In the rest of the book, we will use "fraction of nodes that are Byzantines" and "fraction of Byzantine nodes" interchangeably.

1.4 Outline

The main content of the book is provided in Chaps. 2–6. Chapter 2 introduces the mathematical background of conventional inference by describing the typical inference problems of detection, estimation, classification, and tracking. Specific challenges associated with inference problems in practical sensor networks are discussed, with emphasis on Byzantine attacks. Finally, a taxonomy of results is presented: The results are divided into fundamental limits of attack strategies from the Byzantine's perspective and mitigation strategies from the network's perspective. In each of the subsequent chapters, different inference problems are considered, and results are presented both from attacker's and network designer's perspective under different network architectures.

Chapter 3 discusses networked detection problems for several practical network architectures such as parallel, multi-hop, and fully autonomous ad hoc networks. Following the taxonomy followed throughout the book, fundamental limits of networked inference in the presence of Byzantine attacks are first presented. Next, design of optimal countermeasures (using the insights provided by these fundamental limits) is discussed from a network designer's perspective, and efficient heuristic solutions are provided for cases which are computationally hard.

Chapter 4 discusses the distributed parameter estimation problems with special emphasis on target localization in WSNs. Fundamental limits of distributed estimation with Byzantines are first presented. After characterizing these limits, several countermeasures are discussed, among which some are specific to the estimation framework as they leverage the structure of the parameter space. Specifically, recently developed error-correcting codes-based mitigation scheme is discussed.

Chapter 5 discusses additional inference topics in the presence of Byzantine attacks, such as distributed inference with non-binary quantization and distributed target tracking. These problems raise some unique challenges and are addressed in detail in this chapter.

Chapter 6 presents some recent directions in networked inference with Byzantines. Specifically, the concept of "friendly" Byzantines is discussed where cooperative Byzantines are intentionally introduced in the network to counter the effect of eavesdroppers in the network. Another direction of recent research has focused on using techniques developed for robust networked inference for ensuring reliable crowdsourcing in the presence of unreliable workers. The techniques developed to mitigate the effect of Byzantines in inference problems can be used for crowdsourcing systems, as discussed in this chapter. Finally, a framework called Human–Machine Inference Network (HuMaIN) is introduced, which is a network with both sensors and humans working together to solve networked inference problems. This framework is relevant in the age of IoT where all (exponentially increasing number of) devices are connected with humans in daily life. Some open research problems are also discussed.

References

1. Lamport L, Shostak R, Pease M (1982) The Byzantine generals problem. IEEE Trans Autom Control 4(3):382–401. https://doi.org/10.1145/357172.357176
2. Shannon CE (1956) Reliable machines from unreliable components. MIT, Cambridge, MA (notes of first five lectures in the seminar of information theory)
3. Varshney PK (1996) Distributed detection and data fusion. Springer, New York
4. Veeravalli VV, Varshney PK (2012) Distributed inference in wireless sensor networks. Phil Trans R Soc A 370(1958):100–117. https://doi.org/10.1098/rsta.2011.0194

Chapter 2
Background

In this chapter, we present the relevant background for the material presented in this book. We first present the fundamental concepts of inference in the context of signal processing, including hypothesis testing and estimation techniques. The paradigm of distributed inference in networks is presented along with some associated practical concerns. Finally, the taxonomy of results in this field is outlined.

2.1 Foundational Concepts of Inference

Inferring about an unknown phenomenon based on observations of multiple agents is a problem that we face everyday. Almost every decision we make is made after collecting evidence from multiple sources (agents). These sources are typically of two kinds: sensors/machines that are objective in nature and humans that are typically subjective in nature (sometimes referred to as *hard* and *soft* sources, respectively). Inference is the process of making decisions about a phenomenon using multiple observations and/or evidences regarding it.

The typical inference problems are of two major kinds: classification and estimation [15, 16]. These can be further separated into two sub-types each. In the following, we summarize the different types of inference problems and their general solution methodologies.

2.1.1 Binary Hypothesis Testing

Consider the setup where the goal is to infer (from observations) one among two possible hypotheses. This problem is solved by building models of how the observed evidence is expected to behave under each of the two hypotheses. Based on the

A. Vempaty et al., *Secure Networked Inference with Unreliable Data Sources*,
https://doi.org/10.1007/978-981-13-2312-6_2

observed data and the built models, a *hypothesis test* is performed which lets us decide in favor of one of the hypotheses. In the signal processing literature, this is also referred to as a *detection* problem where one is trying to *detect* the presence or absence of a hypothesis.

Let $\mathbf{x} = [x_1, \ldots, x_N]$ be the N observations collected regarding the phenomenon and let $p(\mathbf{x}|H_l)$ for $l = 0, 1$, be the distribution[1] of data under the two hypotheses H_0 and H_1. Based on the observations and the model, the hypothesis testing problem is typically solved by the maximum *a posteriori* (MAP) test:

$$\frac{p(H_1|\mathbf{x})}{p(H_0|\mathbf{x})} \underset{H_0}{\overset{H_1}{\gtrless}} \lambda, \tag{2.1}$$

which can also be written in terms of the likelihood functions as the following *likelihood-ratio test* (LRT):

$$\frac{p(\mathbf{x}|H_1)}{p(\mathbf{x}|H_0)} \underset{H_0}{\overset{H_1}{\gtrless}} \eta. \tag{2.2}$$

Both thresholds λ and η depend on the problem setup and are determined based on the required objectives of the hypothesis test. The two major solution frameworks are the Neyman–Pearson [19] and Bayesian hypothesis testing frameworks [16].

Depending on the solution framework, the performance of binary hypothesis testing is evaluated in terms of the type-I and type-II errors, which are also referred to as the false alarm and missed detection probabilities respectively. Please refer to [16] for details.

2.1.2 M-Ary Hypothesis Testing

In M-ary hypothesis testing or classification, the phenomenon to be inferred is of finite possibilities, say M possible classes. This is represented by M possible hypotheses: $H_0, \ldots, H_{M-1}$. Multiple observations $\mathbf{x} = [x_1, \ldots, x_N]$ are collected regarding the phenomenon which follow distribution $p(\mathbf{x}|H_l)$ for each hypothesis $H_l, l = 0, \ldots, M-1$. The true class is then determined using the M-ary MAP decision rule as, decide H_k if

$$p(H_k|\mathbf{x}) \geq p(H_i|\mathbf{x}), \forall i \neq k. \tag{2.3}$$

[1] Strictly speaking, it is the probability density function (pdf), but we use the term distribution and pdf interchangeably.

If multiple hypotheses satisfy the condition, ties are broken randomly. For the case of equiprobable prior, this becomes the M-ary maximum likelihood (ML) decision rule. The performance of a typical M-ary hypothesis testing problem is evaluated in terms of the confusion matrix or misclassification probability. Note that the special case when the number of hypotheses $M = 2$, the classification problem reduces to the detection problem discussed above.

2.1.3 Parameter Estimation

In a typical estimation problem, the goal is to estimate the value of an unknown parameter θ (which could possibly be a vector). This value is continuous[2] and can take any value in the set Θ. Similar to the classification (or detection) problem, multiple observations are taken $\mathbf{x} = [x_1, \ldots, x_N]$ that follow a conditional distribution $p(\mathbf{x}|\theta)$. When θ is random, the Bayesian estimate $\hat{\theta}$ is obtained as the MAP estimator

$$\hat{\theta} = \arg\max_{\theta} p(\theta|\mathbf{x}). \tag{2.4}$$

When the prior is uniform, it reduces to the ML estimator. For this setup of random parameter θ following a prior distribution $p(\theta)$, the performance is evaluated using mean square error (MSE) which is the expected squared error between the true parameter value θ and the estimate $\hat{\theta}$:

$$MSE(\theta) = E_{\theta,\hat{\theta}}\left[\left(\hat{\theta} - \theta\right)^2\right], \tag{2.5}$$

where the expectation is taken over the joint distribution of θ and $\hat{\theta}$. Another typically considered estimate is also the minimum mean square error (MMSE) estimate given by

$$\hat{\theta} = \arg\min_{\theta} \text{MSE}(\theta). \tag{2.6}$$

When the unknown parameter θ is deterministic, we use an ML estimator given by

$$\hat{\theta} = \arg\max_{\theta} p(\mathbf{x}|\theta), \tag{2.7}$$

and the performance of the estimation problem is evaluated by reduced MSE, where the expectation in Eq. 2.5 is only with respect to the distribution of $\hat{\theta}$.

[2]We are focusing on the continuous case. Discrete cases can be considered in a similar manner.

2.1.4 Asymptotic Performance Metrics

In this section, we summarize the different asymptotic performance metrics used in the literature for different problem setups. These metrics are used as a measure to quantify the performance in many practical cases when the evaluation of exact performance is intractable.

2.1.4.1 Neyman–Pearson Detection

From Stein's lemma [10], we know that the Kullback–Leibler divergence (KLD) represents the best error exponent of the missed detection error probability in the Neyman–Pearson (NP) setup [19]. For a fixed false alarm probability, $P_F \leq \delta$, the missed detection probability for an optimal NP detector asymptotically behaves as

$$\lim_{N\to\infty} \frac{1}{N} \log P_M = -D\left(H_0||H_1\right), \tag{2.8}$$

where N is the number of samples used for inference and $D(H_0||H_1)$ is the Kullback–Leibler divergence (KLD) between the conditional distributions of data under H_0 and H_1. A direct consequence of this statement is that P_M decays, as N grows to infinity, exponentially, i.e.,

$$P_M \approx f(N)e^{-D(H_0||H_1)}, \tag{2.9}$$

where $f(N)$ is a slow-varying function compared to the exponential, such that $\lim_{N\to\infty} \frac{1}{N} \log f(N) = 0$. Therefore, given a large number of observations, the detection performance depends exclusively on the KLD between the hypotheses. We can conclude that the larger the KLD is, the less is the likelihood of mistaking H_0 with H_1 and therefore, KLD can be used as a surrogate for the probability of missed detection during system design for a large system.

2.1.4.2 Bayesian Detection

While missed detection and false alarm probabilities are individually considered in the NP framework, Bayesian detection considers both of them by weighing them with their prior probability. This overall probability of error is given by

$$P_E = p_0 P_F + p_1 P_M \tag{2.10}$$

where $p_0 = Pr(H_0)$ is the prior probability that the null hypothesis is true, $p_1 = Pr(H_1) = 1 - p_0$ is the prior probability that the alternate hypothesis is true, and P_F and P_M are the false alarm and missed detection probabilities as defined before.

For an optimal detector, this error probability also decays exponentially, and the Chernoff information represents the best error exponent of this error probability in the Bayesian setup:

$$\lim_{N\to\infty} \frac{1}{N} \log P_E = -C\left(H_0, H_1\right), \tag{2.11}$$

where N is the number of samples used for inference and $C(H_0, H_1)$ is the Chernoff information between the conditional distributions of data under H_0 and H_1.

2.1.4.3 Estimation: Deterministic Parameter

For an unknown deterministic parameter, a non-Bayesian estimator is used to give the estimate of the parameter. As indicated earlier, a well-known non-Bayesian estimator is the ML estimator, which maximizes the likelihood $\log p(\mathbf{x}|\theta)$ where $\mathbf{x}$ is the observed data. For any unbiased non-Bayesian estimator, its covariance matrix is bounded by the Cramer–Rao Lower Bound (CRLB) [15]:

$$E\left\{\left[\hat{\theta}-\theta\right]\left[\hat{\theta}-\theta\right]^T\right\} \geq \mathbf{J}^{-1} \tag{2.12}$$

in which the inequality sign is defined in the positive semidefinite sense. $\mathbf{J}$ is the fisher information matrix (FIM), whose inverse provides the CRLB matrix, and

$$\mathbf{J} = E\left[-\Delta_\theta^\theta \log p(\mathbf{x}|\theta)\right] \tag{2.13}$$

where $\Delta_\theta^\theta = \nabla_\theta \nabla_\theta^T$ is the second derivative (Hessian) operator, and ∇ is the gradient operator.

2.1.4.4 Estimation: Random Parameter

For a random parameter with prior $p(\theta)$, two widely used Bayesian estimators are the maximum a posteriori (MAP) estimator, which maximizes the posterior probability density function (PDF), $\arg\max_\theta p(\theta|\mathbf{x})$, and the minimum mean square error (MMSE) estimator, which is also the conditional mean $E\{\theta|\mathbf{x}\}$. For Bayesian estimation problems, the posterior CRLB (PCRLB) [4, 28], or the Bayesian CRLB, provide a lower bound on the estimation mean square error (MSE):

$$E[\hat{\theta}-\theta][\hat{\theta}-\theta]^T \geq \mathbf{F}^{-1} \tag{2.14}$$

where $\mathbf{F}$ is the Bayesian FIM

$$\mathbf{F} = E\left[-\Delta_\theta^\theta \log p(\theta, \mathbf{x})\right]. \tag{2.15}$$

where $\Delta_\theta^\theta = \nabla_\theta \nabla_\theta^T$ is the second derivative (Hessian) operator, and ∇ is the gradient operator.

2.2 Distributed Inference

In the case of distributed inference (classification or estimation), the observations are made in a distributed manner using multiple agents that are spatially distributed in a network, e.g., in a wireless sensor network. These agents observe the phenomenon and transmit their observations over (possibly) imperfect channels to a FC (in a centralized setup) who then fuses the data to infer about the phenomenon. The typical goal in such a framework is to design the signal processing schemes at the local sensors and the FC to infer regarding an event as accurately as possible, subject to an alphabet-size constraint on the messages transmitted by each local sensor (for a comprehensive survey, see [30] and references therein). Several aspects of such a framework can be studied [30]: network topology, decision rules, effect of wireless channels, effect of spatiotemporal dependence, etc. Most of the initial work focused on the design of local sensor decision rules and optimal fusion rule at the FC [5, 9, 14, 17, 18, 23, 27, 29, 31–33]. The advancement of wireless sensor networks (WSNs) renewed interest in this area along with new research challenges: wireless channels [7, 8], network topologies [1, 25, 26], sensor resource management [2, 3, 13, 22], correlated observations [6, 11, 12, 24], etc. The effect of wireless channels was addressed by analyzing the system under the channel-aware formulation [8, 20].

Distributed inference has been extensively studied by various authors in the past few decades. In the context of distributed inference with multiple sensors in a sensor network, a good survey can be found in [30] and references therein. When the agents performing the inference task are humans, this setup is similar to that of team decision making studied by Radner in [21].

2.2.1 Network Topologies

A major segment of distributed inference research has focused on the effect of topologies on system performance. Researchers have considered the move from the traditional parallel topology (also referred to as the star topology [29]) to other topologies such as serial/tandem topologies [25] and tree topologies [26].

Traditional parallel networks comprised of N nodes and a centralized FC. As described above, nodes collect the information simultaneously, carry out local processing, and transmit the processed data directly to the FC where the final inference is made regarding the phenomenon of interest (POI). Note that, nodes carry out local computation/processing independently without collaborating with each other.

Tree or multi-hop networks, with depth K (greater than 1), comprised of N_k nodes at level $k \in \{1, \ldots, K\}$, and a FC at the root (or level 0) of the tree. Observations are acquired and processed by leaf nodes at Level K and sent to their parent nodes, each of which may fuse all the messages it receives with its own measurement if any and then forward the new message to its parent node at the next level. This process takes place throughout the tree, culminating at the root (or FC) where the global inference is made. In this manner, information from each node is aggregated at the FC sent via multi-hop paths.

In decentralized peer-to-peer networks, there is no centralized FC available. The network topology in such scenarios is modeled as a directed graph $G(V, E)$ with $|V| = N$ nodes. The set of communication links in the network correspond to the set of edges E, where an edge exists if and only if there is a communication link between nodes to directly communicate with each other. In order to reach a global inference, peer-to-peer local information exchange schemes, e.g., consensus, gossip algorithm, and diffusion, are employed where each node communicates only with its neighbors according to a pre-specified local fusion rule.

2.2.2 Practical Concerns

Depending on the problem of interest, the agents in distributed inference networks are of two types: physical sensors making objective measurements or human workers providing subjective opinions. Most of the work in the wireless sensor network (WSN) literature has been when these agents are physical sensors. More recently, the crowdsourcing paradigm is considering the case when the distributed agents are human workers performing a distributed inference task. The data transmitted by the local agents to the FC is quantized due to practical constraints such as limited energy or bandwidth in the case of physical sensors and/or cognitive constraints in the case of human agents. The physical sensors are sometimes referred to as 'hard' sensors and the humans are sometimes referred to as 'soft' sensors.

These multiple agents in the distributed inference network are not necessarily reliable. For example, in the case of physical sensors, they could be unreliable due to noisy local observations received at the local sensors. This is governed by the conditional distributions described earlier. Besides this basic cause, there are several other causes for unreliable data from such physical sensors. The sensors could have permanent faults such as stuck-at faults which cause the sensor to always send the same information to the FC irrespective of what it senses. Such a behavior of the sensor provides no information to the FC and therefore needs to be addressed. A more malicious cause could be the case of security attacks where the sensor can be attacked and reprogrammed by an external adversary into sending false data to the FC. Such a malicious sensor sending false information would result in a deteriorated performance at the FC if suitable mitigation approaches are not employed. All these reasons could result in unreliable data at the FC from the physical sensors. In the case of humans, unreliable data could be due to the lack of skill by the human worker that could result in imperfect information from him/her. Although unintentional, the lack of knowledge has the same effect as other unreliable data and can cause a degraded performance at the FC. Sometimes, the unreliableness could also be due to the spammers in the network who provide random data as done in crowdsourcing networks where the workers are typically anonymous. Besides such scenarios, the maliciousness may also exist in some cases, where an external user gets into the system to provide intentional false data while also learning about the inference task.

2.3 Taxonomy of Results

This book considers the problem of distributed inference in the presence of imperfect/unreliable observations and presents results for different inference problems and network topologies. These results follow two major threads: fundamental limits of attack and mitigation techniques against security threats.

2.3.1 Fundamental Limits

For secure networked inference, there are two major players involved: the Byzantines (malicious nodes) and the network (honest nodes). The results on fundamental limits are about learning the limits of attack from attacker's perspective so as to cause maximum damage and the corresponding limit on the performance degradation. More specifically, we derive the minimum attack *power* needed by the Byzantines to cause maximum deterioration of the inference performance. To determine such a limit, one has to characterize the performance associated with the inference problem and determine the degradation in performance as a function of the Byzantine power.

One important result is to determine the minimum fraction α^* of Byzantine sensors that makes the inference at the FC no better than merely making a random inference without the help of any observed data. We call α the attack power of the Byzantine sensors and α^* as the *blinding* power. Besides characterizing the limit, it is also important to consider the case when the power of the adversary is less than this critical value. In such a case, several important questions arise: what should the Byzantines do to cause maximum deterioration in network's performance, what should be the inference rule at the FC, and to what degree is the performance affected?

More precisely, depending on the inference task and the characterization of performance of both asymptotic and non-asymptotic cases as discussed in Sect. 2.1, the *blinding* power α^* and the optimal attack strategies can be derived. In this book, these limits and attack strategies are presented for different inference paradigms.

2.3.2 Mitigation Schemes

Another thread of results presented in this book is from the network's perspective where mitigation schemes are presented to counter the Byzantine attacks. These schemes are of two major types: Byzantine identification schemes and Byzantine tolerant schemes.

In Byzantine identification schemes, the Byzantines in the network are identified based on their data and are either removed from the network or considered a part of the inference after some post-processing of their data. In Byzantine tolerant schemes, inference strategies are developed that are robust to the presence of Byzantines sending falsified data.

In this book, the results follow this taxonomy for different inference paradigms and network topologies: define system and attack models, derive optimal Byzantine attack strategies, and develop mitigation schemes from network's perspective.

References

1. Alhakeem S, Varshney PK (1995) A unified approach to the design of decentralized detection systems. IEEE Trans Aerosp Electron Syst 31(1):9–20. https://doi.org/10.1109/7.366288
2. Appadwedula S, Veeravalli VV, Jones DL (2005) Energy-efficient detection in sensor networks. IEEE J Sel Areas Commun 23(4):693–702. https://doi.org/10.1109/JSAC.2005.843536
3. Appadwedula S, Veeravalli VV, Jones DL (2008) Decentralized detection with censoring sensors. IEEE Trans Signal Process 56(4):1362–1373. https://doi.org/10.1109/TSP.2007.909355
4. Bell KL (1995) Performance bounds in parameter estimation with application to bearing estimation. PhD thesis, George Mason University
5. Blum RS, Kassam SA, Poor HV (1997) Distributed detection with multiple sensors: Part II-Advanced topics. Proc IEEE 85(1):64–79. https://doi.org/10.1109/5.554209
6. Chamberland JF, Veeravalli VV (2006) How dense should a sensor network be for detection with correlated observations? IEEE Trans Inf Theory 52(11):5099–5106. https://doi.org/10.1109/TIT.2006.883551
7. Chamberland JF, Veeravalli VV (2007) Wireless sensors in distributed detection applications. IEEE Signal Process Mag 24(3):16–25. https://doi.org/10.1109/MSP.2007.361598
8. Chen B, Tong L, Varshney PK (2006) Channel-aware distributed detection in wireless sensor networks. IEEE Signal Process Mag 23(4):16–26. https://doi.org/10.1109/MSP.2006.1657814
9. Chen H, Varshney PK (2009) Performance limit for distributed estimation systems with identical one-bit quantizers. IEEE Trans Signal Process 58(1):1607–1621. https://doi.org/10.1109/JPROC.2014.2341554
10. Cover TM, Thomas JA (1991) Elements of information theory. Wiley, New York
11. Dabeer O, Masry E (2008) Multivariate signal parameter estimation under dependent noise from 1-bit dithered quantized data. IEEE Trans Inf Theory 54(4):1637–1654. https://doi.org/10.1109/TIT.2008.917637
12. Iyengar S, Varshney PK, Damarla T (2011) A parametric copula based framework for hypotheses testing using heterogeneous data. IEEE Trans Signal Process 59(5):2308–2319. https://doi.org/10.1109/TSP.2011.2105483
13. Jiang R, Chen B (2005) Fusion of censored decisions in wireless sensor networks. IEEE Trans Wirel Commun 4(6):2668–2673. https://doi.org/10.1109/TWC.2005.858363
14. Kar S, Chen H, Varshney PK (2012) Optimal identical binary quantizer design for distributed estimation. IEEE Trans Signal Process 60(7):3896–3901. https://doi.org/10.1109/TSP.2012.2191777
15. Kay SM (1993) Fundamentals of statistical signal processing: estimation theory. Prentice Hall PTR, Upper Saddle River, NJ
16. Kay SM (1998) Fundamentals of statistical signal processing: detection theory. Prentice Hall PTR, Upper Saddle River, NJ
17. Lam WM, Reibman AR (1993) Design of quantizers for decentralized estimation systems. IEEE Trans Comput 41(11):1602–1605. https://doi.org/10.1109/26.241739
18. Luo ZQ (2005) Universal decentralized estimation in a bandwidth constrained sensor network. IEEE Trans Inf Theory 51(6):2210–2219. https://doi.org/10.1109/TIT.2005.847692
19. Neyman J, Pearson ES (1928) On the use and interpretation of certain test criteria for purposes of statistical inference: Part I. Biometrika 20A(1/2):175–240. https://doi.org/10.2307/2331945

20. Ozdemir O, Niu R, Varshney PK (2009) Channel aware target localization with quantized data in wireless sensor networks. IEEE Trans Signal Process 57(3):1190–1202. https://doi.org/10.1109/TSP.2008.2009893
21. Radner R (1962) Team decision problems. Ann Math Stat 33(3):857–881. http://www.jstor.org/stable/2237863
22. Rago C, Willett P, Bar-Shalom Y (1996) Censoring sensors: a low-communication-rate scheme for distributed detection. IEEE Trans Aerosp Electron Syst 32(2):554–568. https://doi.org/10.1109/7.489500
23. Ribeiro A, Giannakis GB (2006) Bandwidth-constrained distributed estimation for wireless sensor networks-Part I: Gaussian case. IEEE Trans Signal Process 54(3):1131–1143. https://doi.org/10.1109/TSP.2005.863009
24. Sundaresan A, Varshney PK, Rao NSV (2011) Copula-based fusion of correlated decisions. IEEE Trans Aerosp Electron Syst 47(1):454–471. https://doi.org/10.1109/TAES.2011.5705686
25. Swaszek PE (1993) On the performance of serial networks in distributed detection. IEEE Trans Aerosp Electron Syst 29(1):254–260. https://doi.org/10.1109/7.249133
26. Tay WP, Tsitsiklis JN, Win MZ (2008) Data fusion trees for detection: does architecture matter? IEEE Trans Inf Theory 54(9):4155–4168. https://doi.org/10.1109/TIT.2008.928240
27. Tenney RR, Sandell NR Jr (1981) Detection with distributed sensors. IEEE Trans Aerosp Electron Syst AES-17(4):501–510. https://doi.org/10.1109/TAES.1981.309178
28. Van Trees HL, Bell KL (2007) Bayesian bounds for parameter estimation and nonlinear filtering/tracking. Wiley-IEEE, Hoboken, NJ
29. Varshney PK (1996) Distributed detection and data fusion. Springer, New York
30. Veeravalli VV, Varshney PK (2012) Distributed inference in wireless sensor networks. Phil Trans R Soc A 370(1958):100–117. https://doi.org/10.1098/rsta.2011.0194
31. Vempaty A, He H, Chen B, Varshney PK (2014) On quantizer design for distributed Bayesian estimation in sensor networks. IEEE Trans Signal Process 62(20):5359–5369. https://doi.org/10.1109/TSP.2014.23509645
32. Venkitasubramaniam P, Tong L, Swami A (2007) Quantization for maximin ARE in distributed estimation. IEEE Trans Signal Process 55(7):3596–3605. https://doi.org/10.1109/TSP.2007.894279
33. Viswanathan R, Varshney PK (1997) Distributed detection with multiple sensors: Part I-Fundamentals. Proc IEEE 85(1):54–63. https://doi.org/10.1109/5.554208

Chapter 3
Distributed Detection with Unreliable Data Sources

In this chapter, we discuss networked detection problems for several practical network architectures such as parallel, multi-hop, and fully autonomous ad hoc networks. Following the taxonomy presented earlier, fundamental limits of networked inference in the presence of Byzantine attacks are first presented. Next, design of optimal countermeasures using the insights provided by the fundamental limits is discussed from a network designer's perspective.

3.1 Distributed Bayesian Detection with Byzantines: Parallel Networks

3.1.1 System Model

Consider the binary hypothesis testing problem with two hypotheses H_0 (signal is absent) and H_1 (signal is present). Also, consider a parallel network (see Fig. 3.1), comprised of a Fusion Center (FC) and a set of N sensors (nodes), which faces the task of determining which of the two hypotheses is true. Prior probabilities of the two hypotheses H_0 and H_1 are denoted by P_0 and P_1, respectively. The sensors observe the phenomenon, carry out local computations to decide the presence or absence of the phenomenon, and then send their local decisions to the FC that yields a final decision after processing the local decisions. Observations at the nodes are assumed to be conditionally independent and identically distributed given the hypothesis. A Byzantine attack on such a system compromises some of the nodes which may intentionally send falsified local decisions to the FC to make the final decision incorrect. Let us assume that a fraction α of the N nodes which observe the phenomenon have been compromised by an attacker and consider a Clairvoyant case where the FC knows the values of α and $(P_0,\ P_1)$. Based on its own observations, each node i makes a one-bit local decision $v_i \in \{0, 1\}$ regarding the absence or presence of the phenomenon using the likelihood-ratio test

A. Vempaty et al., *Secure Networked Inference with Unreliable Data Sources*,
https://doi.org/10.1007/978-981-13-2312-6_3

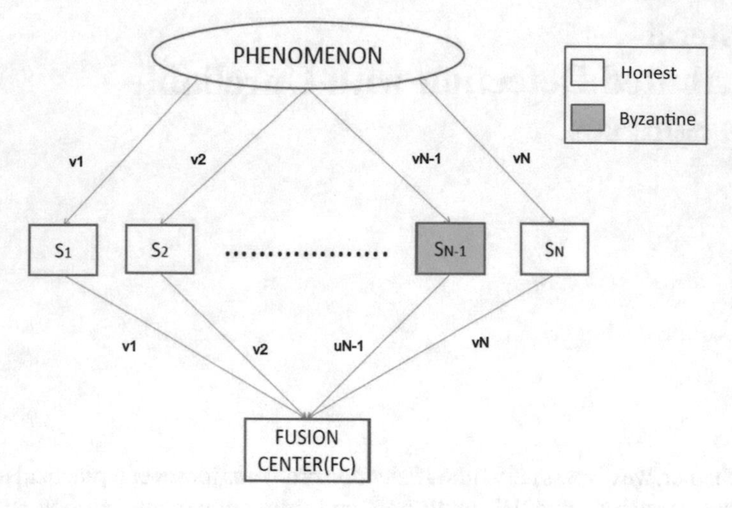

Fig. 3.1 Parallel network architecture

$$\frac{p_{Yi}^{(1)}(y_i)}{p_{Yi}^{(0)}(y_i)} \underset{v_i=0}{\overset{v_i=1}{\gtrless}} \lambda, \tag{3.1}$$

where λ is the identical threshold[1] used at all the sensors and $p_{Yi}^{(k)}(y_i)$ is the conditional probability density function (PDF) of observation y_i under the hypothesis H_k. Each node i, after making its one-bit local decision v_i, sends $u_i \in \{0, 1\}$ to the FC. Let us denote the probabilities of detection and false alarm of each node i in the network by $P_d = P(v_i = 1|H_1)$ and $P_f = P(v_i = 1|H_0)$, respectively.

Binary Hypothesis Testing at the Fusion Center

Consider a Bayesian detection problem where the performance criterion at the FC is the probability of error. The FC receives the decision vector, $\mathbf{u} = [u_1, \cdots, u_N]$, from the nodes and makes the global decision about the phenomenon by considering the maximum a posteriori probability (MAP) rule which is given by

$$P(H_1|\mathbf{u}) \underset{H_0}{\overset{H_1}{\gtrless}} P(H_0|\mathbf{u}) \tag{3.2}$$

or equivalently,

$$\frac{P(\mathbf{u}|H_1)}{P(\mathbf{u}|H_0)} \underset{H_0}{\overset{H_1}{\gtrless}} \frac{P_0}{P_1}. \tag{3.3}$$

Since the u_is are independent of each other, the MAP rule simplifies to a K-out-of-N fusion rule which decides H_1 if K or more u_is are 1 and H_0 otherwise [32]. The global

[1]It has been shown that the use of identical thresholds is asymptotically optimal [31].

false alarm probability Q_F and detection probability Q_D are then given by[2]

$$Q_F = \sum_{i=K}^{N} \binom{N}{i} (\pi_{1,0})^i (1 - \pi_{1,0})^{N-i} \tag{3.4}$$

and

$$Q_D = \sum_{i=K}^{N} \binom{N}{i} (\pi_{1,1})^i (1 - \pi_{1,1})^{N-i}, \tag{3.5}$$

where $\pi_{j,0}$ and $\pi_{j,1}$ are the conditional probabilities of $u_i = j \in \{0, 1\}$ given H_0 and H_1, respectively. The local probability of error at the node as seen by the FC is defined as

$$P_e = P_0 \pi_{1,0} + P_1 \left(1 - \pi_{1,1}\right) \tag{3.6}$$

and the system-wide probability of error at the FC is given by

$$P_E = P_0 Q_F + P_1 (1 - Q_D). \tag{3.7}$$

3.1.2 Analysis from Attacker's Perspective

In [16, 17], the authors considered that each Byzantine decides to attack independently relying on its own observation and derived the fundamental limits and designed mitigation strategies.

Attack Model

For an uncompromised (honest) node i, we have $u_i = v_i$. However, for a compromised (Byzantine) node i, u_i need not be equal to v_i. Let us define the following strategies $P^H_{j,1}, P^H_{j,0}$ and $P^B_{j,1}, P^B_{j,0}$ ($j \in \{0, 1\}$) for the honest and Byzantine nodes, respectively:

Honest nodes:

$$P^H_{1,1} = 1 - P^H_{0,1} = P^H(x = 1|y = 1) = 1 \tag{3.8}$$

$$P^H_{1,0} = 1 - P^H_{0,0} = P^H(x = 1|y = 0) = 0 \tag{3.9}$$

Byzantine nodes:

$$P^B_{1,1} = 1 - P^B_{0,1} = P^B(x = 1|y = 1) \tag{3.10}$$

$$P^B_{1,0} = 1 - P^B_{0,0} = P^B(x = 1|y = 0) \tag{3.11}$$

[2]These expressions are valid under the assumption that $\alpha < 0.5$. Later in Sect. 3.1.2, we will discuss a generalization of results for any arbitrary α.

$P^H(x = a|y = b)$ ($P^B(x = a|y = b)$) is the probability that an honest (Byzantine) node sends a to the FC when its actual local decision is b. From now on, we refer to Byzantine flipping probabilities simply by $(P_{1,0}, P_{0,1})$. Further, assume that the FC is not aware of the exact set of Byzantine nodes and considers each node i to be Byzantine with a certain probability α; thus, $\pi_{1,0}$ and $\pi_{1,1}$ can be calculated as

$$\pi_{1,0} = \alpha(P_{1,0}(1 - P_f) + (1 - P_{0,1})P_f) + (1 - \alpha)P_f \tag{3.12}$$

and

$$\pi_{1,1} = \alpha(P_{1,0}(1 - P_d) + (1 - P_{0,1})P_d) + (1 - \alpha)P_d, \tag{3.13}$$

where α is the fraction of Byzantine nodes.

Blinding the Fusion Center

The goal of Byzantines is to degrade the detection performance of the FC as much as possible. Let us denote the minimum fraction of Byzantine nodes needed to make the FC "blind" and denote it by α_{blind}. The FC is blind if an adversary can falsify the data that the FC receives from the nodes such that no information is conveyed. In other words, the optimal detector at the FC cannot perform better than simply making the decision based on priors. A closed-form expression for α_{blind} in a distributed Bayesian detection scenario was derived in [17] that is given in Lemma 3.1.

Lemma 3.1 ([17]) *In distributed Bayesian detection, the minimum fraction of Byzantines needed to make the fusion center blind is* $\alpha_{blind} = 0.5$.

Proof The FC is "blind" if the received data **u** does not provide any information about the hypotheses to the FC. The condition to make the FC blind can be stated as

$$P(H_i|\mathbf{u}) = P(H_i) \text{ for } i = 0, 1. \tag{3.14}$$

Applying Bayes' theorem, it can be seen that (3.14) is equivalent to $P(\mathbf{u}|H_i) = P(\mathbf{u})$. Thus, the FC becomes blind if the probability of receiving a given vector **u** is independent of the hypothesis present. In such a scenario, the best that the FC can do is to make decisions solely based on the priors, resulting in the most degraded performance at the FC. Now, using the conditional i.i.d. assumption, under which observations at the nodes are conditionally independent and identically distributed given the hypothesis, condition (3.14) to make the FC blind becomes $\pi_{1,1} = \pi_{1,0}$. This is true only when

$$\alpha[P_{1,0}(P_f - P_d) + (1 - P_{0,1})(P_d - P_f)] + (1 - \alpha)(P_d - P_f) = 0. \tag{3.15}$$

Hence, the FC becomes blind if

$$\alpha = \frac{1}{(P_{1,0} + P_{0,1})}. \tag{3.16}$$

α in (3.16) is minimized when $P_{1,0}$ and $P_{0,1}$ both take their largest values, i.e., $P_{1,0} = P_{0,1} = 1$. Hence, $\alpha_{blind} = 0.5$. □

Note that in most practical systems, we have $\alpha << 0.5$. In such scenarios, the authors in [16, 17] further considered the problem of determining optimal attack strategies to degrade the detection performance. For analytical tractability, the authors in [16] adopted Chernoff information as the performance metric and studied the performance of a distributed Bayesian detection system with Byzantines in the asymptotic regime. The results are summarized in Theorem 3.1.

Theorem 3.1 ([17]) *Optimal attack strategies, $(P^*_{1,0}, P^*_{0,1})$, which minimize the Chernoff information are*

$$(P^*_{1,0}, P^*_{0,1}) = \begin{cases} (p_{1,0}, p_{0,1}) \text{ if } \alpha \geq 0.5 \\ (1, 1) \text{ if } \alpha < 0.5 \end{cases}, \tag{3.17}$$

where $(p_{1,0}, p_{0,1})$ satisfy $\alpha(p_{1,0} + p_{0,1}) = 1$.

The system-wide probability of error P_E is a function of the parameter K of the K-out-of-N fusion rule, which is under the control of the FC, and the parameters $(\alpha, P_{j,0}, P_{j,1})$ that are under the control of the attacker. The FC and the Byzantines may or may not have knowledge of their opponent's strategy. In practice, the FC and the Byzantines will optimize their utility by choosing their actions based on the knowledge of their opponent's behavior. This motivated [17] to address the following question: what are the optimal attack/defense strategies given the knowledge of the opponent's strategies? Study of these practically motivated questions requires non-asymptotic analysis, which was systematically studied under several different scenarios (shown in Table 3.1) by the authors in [17].

Next, let us first discuss how the Byzantines can launch an attack optimally considering that the parameter (K) is under the control of the FC. By assuming error probability to be the performance metric, [17] analyzed the system performance in the non-asymptotic regime.

Optimal Attack Strategies Without the Knowledge of the Fusion Rule

In practice, the Byzantine attacker may not have the knowledge about the fusion rule, i.e., the value of K, used by the FC. In such scenarios, the optimal attack strategy for

Table 3.1 Different scenarios based on the knowledge of the opponent's strategies

Cases	Attacker has the knowledge of the FC's strategies	FC has the knowledge of attacker's strategies
Case 1	No	No
Case 2	Yes	No
Case 3	Yes	Yes
Case 4	No	Yes

Byzantines can be obtained by maximizing the local probability of error at the node as seen by the FC, which is independent of the fusion rule K. The problem can be stated formally as

$$\begin{aligned} \underset{P_{1,0},P_{0,1}}{\text{maximize}} \quad & P_0\pi_{1,0} + P_1(1-\pi_{1,1}) \\ \text{subject to} \quad & 0 \leq P_{1,0} \leq 1 \\ & 0 \leq P_{0,1} \leq 1 \end{aligned} \tag{P3.1}$$

To solve the problem, the authors in [17] analyzed the properties of the objective function, $P_e = P_0\pi_{1,0} + P_1(1-\pi_{1,1})$, with respect to $(P_{1,0}, P_{0,1})$. Note that

$$\frac{dP_e}{dP_{1,0}} = P_0\alpha(1-P_f) - P_1\alpha(1-P_d) \tag{3.18}$$

and

$$\frac{dP_e}{dP_{0,1}} = -P_0\alpha P_f + P_1\alpha P_d. \tag{3.19}$$

By utilizing the monotonicity properties of the objective function with respect to $P_{1,0}$ and $P_{0,1}$ (3.18) and (3.19), the solution of Problem P3.1 can be obtained that is presented in Table 3.2. Note that when $\frac{P_d}{P_f} < \frac{P_0}{P_1} < \frac{1-P_d}{1-P_f}$, both (3.18) and (3.19) are less than zero. P_e then becomes a strictly decreasing function of $P_{1,0}$ as well as $P_{0,1}$. Hence, to maximize P_e, the attacker needs to choose $(P_{1,0}, P_{0,1}) = (0, 0)$. However, the condition $\frac{P_d}{P_f} < \frac{P_0}{P_1} < \frac{1-P_d}{1-P_f}$ holds iff $P_d < P_f$ and, therefore, is not admissible. Similar arguments lead to the rest of the results given in Table 3.2. Note that if there is an equality in the conditions mentioned in Table 3.2, then the solution will not be unique. For example, $\left(\frac{dP_e}{dP_{0,1}} = 0\right) \Leftrightarrow \left(\frac{P_0}{P_1} = \frac{1-P_d}{1-P_f}\right)$ implies that P_e is constant as a function of $P_{0,1}$. In other words, the attacker will be indifferent in choosing the parameter $P_{0,1}$ because any value of $P_{0,1}$ will result in the same probability of error.

Numerical Results

Figure 3.2a shows the local probability of error P_e as a function of $(P_{1,0}, P_{0,1})$ when $(P_0 = P_1 = 0.5)$. It is assumed that the local probability of detection is $P_d = 0.8$ and the local probability of false alarm is $P_f = 0.1$ such that $\frac{P_d}{P_f} = 8$, $\frac{1-P_d}{1-P_f} = 0.2222$, and

Table 3.2 Solution of maximizing local error P_e problem (P1)

Condition	$P_{1,0}$	$P_{0,1}$
$\frac{P_d}{P_f} < \frac{P_0}{P_1} < \frac{1-P_d}{1-P_f}$	0	0
$\frac{P_d}{P_f} > \frac{P_0}{P_1} < \frac{1-P_d}{1-P_f}$	0	1
$\frac{P_d}{P_f} < \frac{P_0}{P_1} > \frac{1-P_d}{1-P_f}$	1	0
$\frac{P_d}{P_f} > \frac{P_0}{P_1} > \frac{1-P_d}{1-P_f}$	1	1

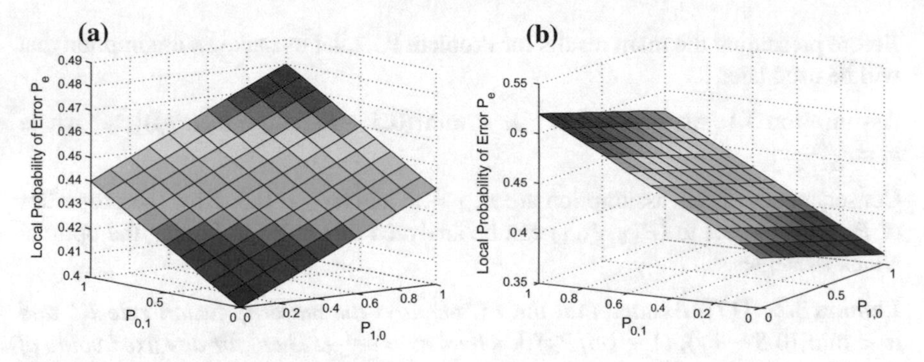

Fig. 3.2 **a** P_e as a function of $(P_{1,0}, P_{0,1})$ when $P_0 = P_1 = 0.5$. **b** P_e as a function of $(P_{1,0}, P_{0,1})$ when $P_0 = 0.1, P_1 = 0.9$ [17]

$\frac{P_0}{P_1} = 1$. Clearly, $\frac{P_d}{P_f} > \frac{P_0}{P_1} > \frac{1-P_d}{1-P_f}$ and it implies that the optimal attack strategy is $(P_{1,0}, P_{0,1}) = (1, 1)$, which can be verified from Fig. 3.2a.

Figure 3.2b presents the local probability of error P_e as a function of the attack strategy $(P_{1,0}, P_{0,1})$ when $(P_0 = 0.1, P_1 = 0.9)$. It is assumed that the local probability of detection is $P_d = 0.8$ and the local probability of false alarm is $P_f = 0.1$ such that $\frac{P_d}{P_f} = 8$, $\frac{1-P_d}{1-P_f} = 0.2222$, and $\frac{P_0}{P_1} = .1111$. Clearly, $\frac{P_d}{P_f} > \frac{P_0}{P_1} < \frac{1-P_d}{1-P_f}$ which implies that the optimal attack strategy is $(P_{1,0}, P_{0,1}) = (0, 1)$, which can be verified from Fig. 3.2b.

Optimal Byzantine Attack Strategies Under Majority Fusion Rule

Next, the authors in [17] investigated the scenario where Byzantines are aware of the fusion rule K used at the FC and can use this knowledge to provide false information in an optimal manner to blind the FC. However, it was assumed that the FC does not have knowledge of Byzantine's attack strategies $(\alpha, P_{j,0}, P_{j,1})$ and does not optimize against Byzantine's behavior. Since majority rule is a widely used fusion rule [11, 26, 37], it was assumed that the FC uses the majority rule to make the global decision. The performance criterion at the FC was assumed to be the probability of error P_E.

For a fixed fusion rule K^*, which, as mentioned before, is assumed to be the majority rule $K^* = \lceil \frac{N+1}{2} \rceil$, P_E varies with the parameters $(\alpha, P_{j,0}, P_{j,1})$ which are under the control of the attacker. The Byzantine attack problem can be formally stated as follows:

$$\begin{aligned} \underset{P_{j,0}, P_{j,1}}{\text{maximize}} \quad & P_E(\alpha, P_{j,0}, P_{j,1}) \\ \text{subject to} \quad & 0 \leq P_{j,0} \leq 1 \\ & 0 \leq P_{j,1} \leq 1. \end{aligned} \tag{P3.2}$$

For a fixed fraction of Byzantines α, the attacker wants to maximize the probability of error P_E by choosing its attack strategy $(P_{j,0}, P_{j,1})$ optimally. Let us assume that the attacker is aware that the FC is using the majority rule for making the global decision.

Before presenting the main results for Problem P3.2, let us make an assumption that will be used later.

Assumption 3.1 Assume that $\alpha < \min\{(0.5 - P_f), (1 - (m/P_d))\}$,[3] where $m = \frac{N}{2N-2}$.

Consequences of this assumption are $\pi_{1,1} > m$ and $\pi_{1,0} < 0.5$. Now the properties of P_E with respect to $(P_{1,0}, P_{0,1})$ can be analyzed that helps in finding the optimal attack strategies.

Lemma 3.2 ([17]) *Assume that the FC employs the majority fusion rule K^* and $\alpha < \min\{(0.5 - P_f), (1 - (m/P_d))\}$, where $m = \frac{N}{2N-2}$. Then, for any fixed value of $P_{0,1}$, the error probability P_E at the FC is a quasi-convex function of $P_{1,0}$.*

Proof The proof is given in Appendix A.1. □

Quasi-convexity of P_E over $P_{1,0}$ implies that the maximum of the function occurs on the corners, i.e., $P_{1,0} = 0$ or 1 (may not be unique). Next, let us analyze the properties of P_E with respect to $P_{0,1}$.

Lemma 3.3 ([17]) *Assume that the FC employs the majority fusion rule K^* and $\alpha < \min\{(0.5 - P_f), (1 - (m/P_d))\}$, where $m = \frac{N}{2N-2}$. Then, the probability of error P_E at the FC is a quasi-convex function of $P_{0,1}$ for a fixed $P_{1,0}$.*

Proof The proof is given in Appendix A.2. □

Theorem 3.2 ([17]) *When the majority fusion rule is employed at the FC and $\alpha < \min\{(0.5 - P_f), (1 - (m/P_d))\}$, where $m = \frac{N}{2N-2}$, we have* (1, 0), (0, 1), *or* (1, 1) *as the optimal attack strategies $(P_{1,0}, P_{0,1})$ that maximize the probability of error P_E.*

Proof Lemmas 3.2 and 3.3 suggest that one of the corners is the maximum of P_E because of quasi-convexity. Note that (0, 0) cannot be the solution of the maximization problem since the attacker does not flip any results. Hence, we end up with three possibilities: (1, 0), (0, 1), or (1, 1). □

Numerical Results

Figure 3.3a shows the probability of error P_E as a function of the attack strategy $(P_{1,0}, P_{0,1})$ for even number of nodes, $N = 10$, in the network. Let us assume that the probability of detection is $P_d = 0.8$, the probability of false alarm is $P_f = 0.1$, prior probabilities are $(P_0 = 0.4, P_1 = 0.6)$, and $\alpha = 0.37$. Since $\alpha < \min\{(0.5 - P_f), (1 - (m/P_d))\}$, where $m = \frac{N}{2N-2}$, quasi-convexity can be observed in Fig. 3.3a. Figure 3.3b shows the probability of error P_E as a function of attack strategy $(P_{1,0}, P_{0,1})$ for odd number of nodes, $N = 11$, in the network. Similarly, quasi-convexity can be observed in Fig. 3.3b.

[3]Condition $\alpha < \min\{(0.5 - P_f), (1 - (m/P_d))\}$, where $m = \frac{N}{2N-2} > 0.5$, suggests that as N tends to infinity, $m = \dfrac{N}{2N-2}$ tends to 0.5. When P_d tends to 1 and P_f tends to 0, the condition $\alpha < \min\{(0.5 - P_f), (1 - (m/P_d))\}$ simplifies to $\alpha < 0.5$.

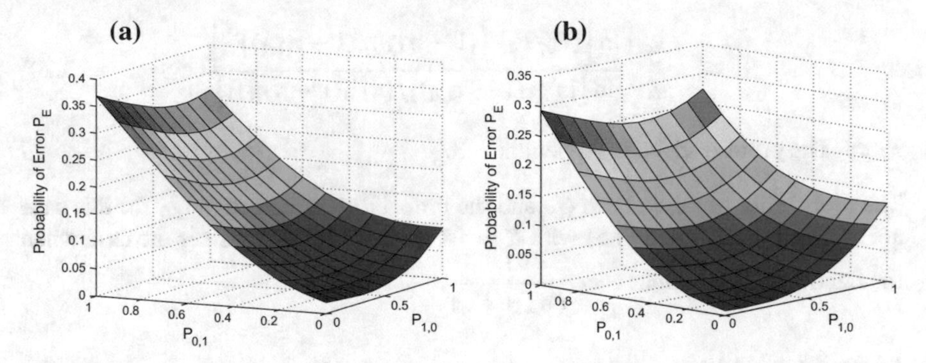

Fig. 3.3 **a** P_E as a function of $(P_{1,0}, P_{0,1})$ for $N = 10$. **b** P_E as a function of $(P_{1,0}, P_{0,1})$ for $N = 11$ [17]

It is evident from Fig. 3.3a and b that the optimal attack strategy $(P_{1,0}, P_{0,1})$ is either of the following three possibilities: (1, 0), (0, 1), or (1, 1).

Observe that the results obtained for this case are not the same as the results obtained for the asymptotic case (see Theorem 3.1). This is because the asymptotic performance measure, i.e., Chernoff information, is the exponential decay rate of the error probability of the "optimal detector." In other words, while optimizing over Chernoff information, it is implicitly assumed that the optimal fusion rule is used at the FC which is not the case here.

Optimal Byzantine Attack Strategies with Strategy-aware FC

Next the authors in [17] investigated the case where the FC has the knowledge of attacker's strategies and uses the optimal fusion rule K^* to make the global decision. Here, the attacker tries to maximize the worst case probability of error $\underset{K}{min} P_E$ by choosing $(P_{1,0}, P_{0,1})$ optimally. The Byzantine attack problem for such a scenario can be formally stated as follows:

$$\begin{aligned} \underset{P_{j,0}, P_{j,1}}{\text{maximize}} \quad & P_E(K^*, \alpha, P_{j,0}, P_{j,1}) \\ \text{subject to} \quad & 0 \le P_{j,0} \le 1 \\ & 0 \le P_{j,1} \le 1, \end{aligned} \tag{P3.3}$$

where K^* is the optimal fusion rule. In other words, K^* is the best response of the FC to the Byzantine attack strategies.

The optimal fusion rule assuming that the local sensor threshold λ and the Byzantine attack strategy $(\alpha, P_{1,0}, P_{0,1})$ are fixed and known to the FC is given as follows.

Lemma 3.4 ([17]) *For a fixed local sensor threshold* λ *and* $\alpha < \dfrac{1}{P_{0,1} + P_{1,0}}$, *the optimal fusion rule is given by*

$$K^* \underset{H_0}{\overset{H_1}{\gtrless}} \frac{\ln\left[(P_0/P_1)\left\{(1-\pi_{1,0})/(1-\pi_{1,1})\right\}^N\right]}{\ln\left[\{\pi_{1,1}(1-\pi_{1,0})\}/\{\pi_{1,0}(1-\pi_{1,1})\}\right]}. \tag{3.20}$$

Proof The proof is given in Appendix A.3. □

The probability of false alarm Q_F and the probability of detection Q_D for this case are as given in (3.4) and (3.5) with $K = \lceil K^* \rceil$. Next, let us consider the case when the fraction of Byzantines $\alpha > \dfrac{1}{P_{0,1}+P_{1,0}}$.

Lemma 3.5 ([17]) *For a fixed local sensor threshold* λ *and* $\alpha > \dfrac{1}{P_{0,1}+P_{1,0}}$, *the optimal fusion rule is given by*

$$K^* \underset{H_1}{\overset{H_0}{\gtrless}} \frac{\ln\left[(P_1/P_0)\left\{(1-\pi_{1,1})/(1-\pi_{1,0})\right\}^N\right]}{[\ln(\pi_{1,0}/\pi_{1,1})+\ln((1-\pi_{1,1})/(1-\pi_{1,0}))]}. \tag{3.21}$$

Proof This can be proved in a similar manner as Lemma 3.4 and using the fact that, for $\pi_{1,1} < \pi_{1,0}$ or equivalently, $\alpha > \dfrac{1}{P_{0,1}+P_{1,0}}$, $[\ln(\pi_{1,0}/\pi_{1,1})+\ln((1-\pi_{1,1})/(1-\pi_{1,0}))] > 0$. □

The probability of false alarm Q_F and the probability of detection Q_D for this case can be calculated to be

$$Q_F = \sum_{i=0}^{\lfloor K^* \rfloor} \binom{N}{i} (\pi_{1,0})^i (1-\pi_{1,0})^{N-i} \tag{3.22}$$

and

$$Q_D = \sum_{i=0}^{\lfloor K^* \rfloor} \binom{N}{i} (\pi_{1,1})^i (1-\pi_{1,1})^{N-i}. \tag{3.23}$$

Now the properties of P_E with respect to Byzantine attack strategy $(P_{1,0}, P_{0,1})$ can be analyzed that helps in finding the optimal attack strategies.

Lemma 3.6 ([17]) *For a fixed local sensor threshold* λ, *assume that the FC employs the optimal fusion rule* $\lceil K^* \rceil$,[4] *as given in (3.20). Then, for* $\alpha \leq 0.5$, *the error probability* P_E *at the FC is a monotonically increasing function of* $P_{1,0}$ *while* $P_{0,1}$ *remains fixed. Conversely, the error probability* P_E *at the FC is a monotonically increasing function of* $P_{0,1}$ *while* $P_{1,0}$ *remains fixed.*

[4] Note that K^* might not be an integer.

Proof The proof is given in Appendix A.5. □

Based on Lemma 3.6, one can derive the optimal attack strategies for the case when the FC has the knowledge regarding the strategies used by the Byzantines.

Theorem 3.3 ([17]) *The optimal attack strategies, $(P^*_{1,0}, P^*_{0,1})$, which maximize the probability of error, $P_E(\lceil K^* \rceil)$, are given by*

$$(P^*_{1,0}, P^*_{0,1}) = \begin{cases} (p_{1,0}, p_{0,1}) \text{ if } \alpha > 0.5 \\ (1, 1) \text{ if } \alpha \leq 0.5 \end{cases} \tag{3.24}$$

where $(p_{1,0}, p_{0,1})$ satisfies $\alpha(p_{1,0} + p_{0,1}) = 1$.

Proof Note that the maximum probability of error occurs when the posterior probabilities are equal to the prior probabilities of the hypotheses. That is,

$$P(H_i|\mathbf{u}) = P(H_i) \text{ for } i = 0, 1. \tag{3.25}$$

The condition can further be simplified to

$$\alpha(P_{1,0} + P_{0,1}) = 1. \tag{3.26}$$

Equation (3.26) suggests that when $\alpha \geq 0.5$, the attacker can find flipping probabilities that make $P_E = \min\{P_0, P_1\}$. When $\alpha = 0.5$, $P_{1,0} = P_{0,1} = 1$ is the optimal attack strategy and when $\alpha > 0.5$, any pair which satisfies $P_{1,0} + P_{0,1} = \dfrac{1}{\alpha}$ is optimal. However, when $\alpha < 0.5$, (3.26) cannot be satisfied. In this case, by Lemma 3.6, for $\alpha < 0.5$, $(1, 1)$ is an optimal attack strategy, $(P_{1,0}, P_{0,1})$, which maximizes the probability of error, $P_E(\lceil K^* \rceil)$. □

Numerical Results

Figure 3.4 shows the minimum probability of error as a function of attacker's strategy $(P_{1,0}, P_{0,1})$, where P_E is minimized over all possible fusion rules K. We consider a $N = 11$ node network, with the detection and false alarm probabilities of each node being 0.6 and 0.4, respectively. Prior probabilities are assumed to be $P_0 = 0.4$ and $P_1 = 0.6$.

Observe that the optimal fusion rule as given in (3.20) changes with attacker's strategy $(P_{1,0}, P_{0,1})$. Thus, the minimum probability of error $\min_K P_E$ is a non-differentiable function. It is evident from Fig. 3.4a that $(P_{1,0}, P_{0,1}) = (1, 1)$ maximizes the probability of error, $P_E(\lceil K^* \rceil)$. This corroborates the theoretical results presented in Theorem 3.3, that for $\alpha < 0.5$, the optimal attack strategy, $(P_{1,0}, P_{0,1})$, that maximizes the probability of error, $P_E(\lceil K^* \rceil)$, is $(1, 1)$.

Figure 3.4b considers the scenario where $\alpha = 0.8$ (i.e., $\alpha > 0.5$). It can be seen that the attack strategy $(P_{1,0}, P_{0,1})$, that maximizes $\min_K P_E$ is not unique in this case. It can be verified that any attack strategy which satisfies $P_{1,0} + P_{0,1} = \frac{1}{0.8}$ will make $\min_K P_E = \min\{P_0, P_1\} = 0.4$. This corroborates the theoretical results presented in

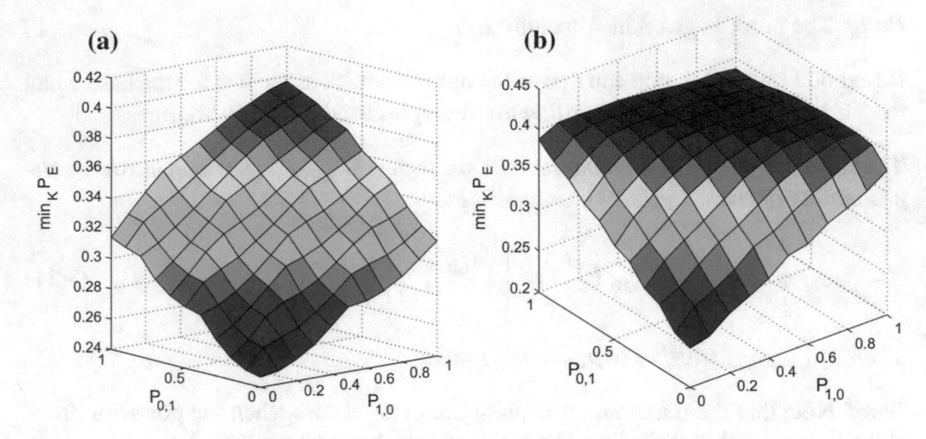

Fig. 3.4 Minimum probability of error ($\min_K P_E$) analysis. **a** $\min_K P_E$ as a function of $(P_{1,0}, P_{0,1})$ for $\alpha = 0.4$. **b** $\min_K P_E$ as a function of $(P_{1,0}, P_{0,1})$ for $\alpha = 0.8$ [17]

Theorem 3.3. Observe that the results obtained for this case are consistent with the results obtained for the asymptotic case. This is because the optimal fusion rule is used at the FC and the asymptotic performance measure, i.e., Chernoff information, is the exponential decay rate of error probability of the "optimal detector" and, thus, implicitly assumes that the optimal fusion rule is used at the FC.

When the attacker does not have the knowledge of the fusion rule K used at the FC, from an attacker's perspective, maximizing its local probability of error P_e is the optimal attack strategy. The optimal attack strategy in this case is either of the three possibilities: $(P_{1,0}, P_{0,1}) = (0, 1)$ or $(1, 0)$ or $(1, 1)$ (see Table 3.2). However, the FC has the knowledge of the attack strategy $(\alpha, P_{1,0}, P_{0,1})$ and, thus, uses the optimal fusion rule as given in (3.20) and (3.21).

3.1.3 Analysis from Network Designer's Perspective

So far, the optimal attack strategies were analyzed for the attacker who intends to degrade the detection performance at the FC. It was shown that when the fraction of Byzantines in the network is greater than 0.5 and is attacking independently, the FC becomes blind to the data from the nodes and can only detect the presence of the phenomenon using the prior information. However, addressing the problem from the network's perspective, one can develop techniques to counter the effect of Byzantines and ensure reliable detection performance. We now explore such a Byzantine mitigation scheme.

Adaptive Learning-Based Byzantine Mitigation Scheme

An effective method to improve the performance of the network is to use the information of the identified Byzantines to the network's benefit [33]. More specifically, a three-tier adaptive learning scheme was proposed in [33] which learns the parameters of the identified Byzantines and uses these learned parameters in the Chair-Varshney rule [3] to make the final decision. The three-tier scheme can be described as follows: (i) identification of Byzantines in the network, (ii) estimation of parameters of the identified Byzantines at the FC, and (iii) adaptive fusion rule.

The basic idea of the learning based identification scheme is to compare every node's observed behavior over time with the expected behavior of an honest node. The nodes whose observed behavior is too far from the expected behavior are tagged as Byzantines. This scheme works even when the Byzantines are in majority (>50%) since it does not use the global decision for identification purposes. The behavior of every node is characterized by the probability of sending a "1" to the FC. This value is a function of the operating point of the node (P_f, P_d) and the prior probabilities of the hypotheses which are assumed to be known at the FC for honest nodes.

At the FC, the expected behavior is estimated for every node over time by averaging the number of times a particular decision (0 or 1) is made over a time interval of T instants. These probabilities can be updated after every time instant. The test statistic Λ_i^T for the ith node after time T is the deviation between the expected and observed behavior for every node. The FC declares a node as a Byzantine if Λ_i^T is greater than a particular threshold λ. This threshold λ is determined as the minimum value when the Byzantine's operating point is in the region below the $P_d = P_f$ line on the receiver operating characteristics (ROC).

After identifying the Byzantines, their parameters can be estimated by assuming that all the Byzantines have the same operating point. This assumption is typically made in the literature since it is assumed that a single adversary has captured some of the nodes in the network and re-programs them to behave as Byzantines. Therefore, it can be assumed that all these malicious nodes have the same operating point on the ROC. These estimated parameters are used in the Chair-Varshney optimal fusion rule [3] in an adaptive manner to find the global decision. It is important to note that this scheme works for any fraction of Byzantines in the network but assumes the knowledge of honest nodes behavior and the signal statistics.

Numerical Results

Consider a 20 node parallel network facing a task of determining the presence (or the absence) of a signal where the prior probability $P_1 = 0.3$. The process of detecting the Byzantines and estimating the probabilities was performed and results can be seen in Fig. 3.5. As can be seen, the Byzantines can be exactly detected without any mismatches when T is between 100 and 150 for both the cases when $\alpha = 0.3$ and $\alpha = 0.7$. It can also be observed that this algorithm works even when the Byzantines are in majority unlike reputation-based scheme proposed in [25].

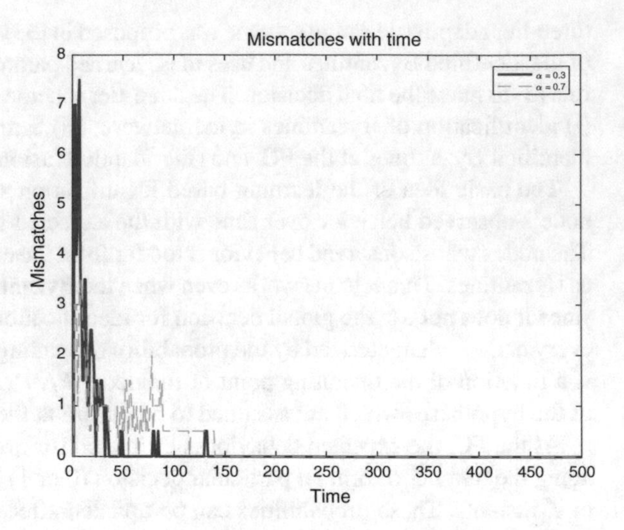

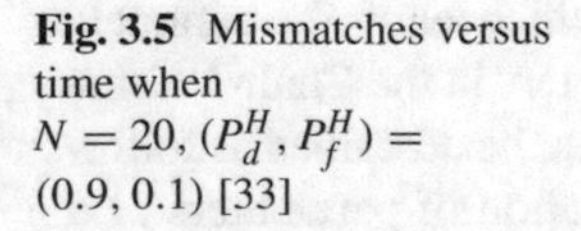

Fig. 3.5 Mismatches versus time when $N = 20$, $(P_d^H, P_f^H) = (0.9, 0.1)$ [33]

3.2 Distributed Neyman–Pearson Detection with Byzantine Data: Parallel Networks

This section discusses distributed detection in a Neyman–Pearson setup with parallel networks.

3.2.1 System Model

Consider the same binary hypothesis testing problem and system model as described in Sect. 3.1.1 and Fig. 3.1. Denote the probabilities of detection and false alarm as $P_d^H = P(u_i = 1|H_1)$ and $P_f^H = P(u_i = 1|H_0)$ for honest nodes and $P_d^B = P(u_i = 1|H_1)$ and $P_f^B = P(u_i = 1|H_0)$ for Byzantines.

3.2.2 Analysis from Attacker's Perspective

In [25], the authors employed the same attack model as given in Sect. 3.1.2 and derived the fundamental limits as well as designed mitigation strategies.

Blinding the Fusion Center

The objective of Byzantines is to degrade the detection performance of the network by choosing $(P_{1,0}, P_{0,1})$ intelligently. Due to the difficulty in characterizing the exact performance of a general fusion scheme, the authors in [25] used Kullback–Leibler divergence (KLD) to characterize the performance of the detection scheme due to its strong relationship with the global detection performance. The KLD between the distributions $P(u_i|H_1)$ and $P(u_i|H_0)$ can be expressed as

$$D(P(u_i|H_1)||P(u_i|H_0)) = \sum_{j\in\{0,1\}} P(u_i|H_1)\ \log \frac{P(u_i|H_1)}{P(u_i|H_0)}. \tag{3.27}$$

where we have

$$P(u_i = j|H_0) = (1-\alpha)[P_f^H P_{j,1}^H + (1-P_f^H)P_{j,0}^H] + \alpha[P_f^B P_{j,1}^B + (1-P_f^B)P_{j,0}^B] \tag{3.28}$$

and

$$P(u_i = j|H_1) = (1-\alpha)[P_d^H P_{j,1}^H + (1-P_d^H)P_{j,0}^H] + \alpha[P_d^B P_{j,1}^B + (1-P_d^B)P_{j,0}^B] \tag{3.29}$$

The critical power of the distributed detection network is the minimum fraction of Byzantines needed to make the data from nodes uninformative to the network, i.e., make KLD at the FC equal to zero. This would result in the FC making a decision based only on the prior information as the network is *blind* to the data from nodes. A closed-form expression for critical power (denoted as α_{blind}) in a distributed Neyman–Pearson detection scenario was derived in [25].

Lemma 3.7 ([25]) *The minimum fraction of Byzantines needed to blind the fusion center is*

$$\alpha_{blind} = \frac{P_d^H - P_f^H}{(P_d^B - P_f^B) + (P_d^H - P_f^H)}. \tag{3.30}$$

Note that when the Byzantine attackers have perfect knowledge of the state of nature, i.e., whether H_0 or H_1 is present ($P_d^B = 1, P_f^B = 0$), (3.30) reduces to the expression derived by Marano et al. in [22].

The optimal attack strategy for the Byzantines was further derived and given by the following.

Lemma 3.8 ([25]) *Let us consider $P_d^B = P_d^H$ and $P_f^B = P_f^H$. Optimal attack strategies, $(P_{1,0}^*, P_{0,1}^*)$, which minimize the KLD are*

$$(P_{1,0}^*, P_{0,1}^*) = \begin{cases} (p_{1,0}, p_{0,1})\ \textit{if}\ \alpha \geq 0.5 \\ (1,1)\ \textit{if}\ \alpha < 0.5 \end{cases}, \tag{3.31}$$

where $(p_{1,0}, p_{0,1})$ satisfy $\alpha(p_{1,0} + p_{0,1}) = 1$.

Numerical Results

To corroborate the theoretical results, Fig. 3.6 shows the detection performance in terms of the minimum KLD[5] as a function of the fraction of Byzantines for different

[5]Minimization is performed over attack strategies $P_{0,1}$ and $P_{1,0}$.

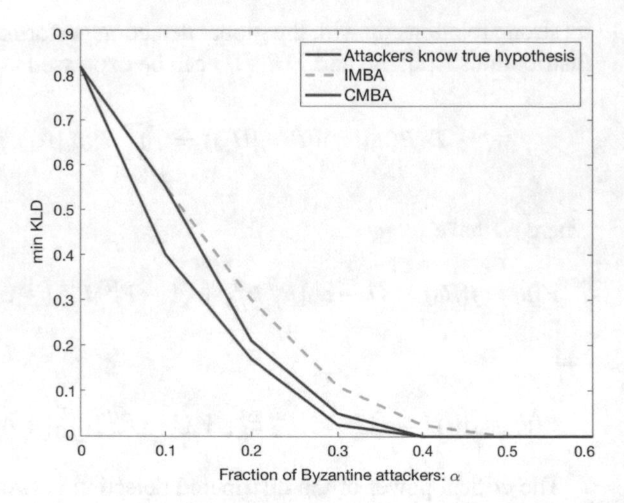

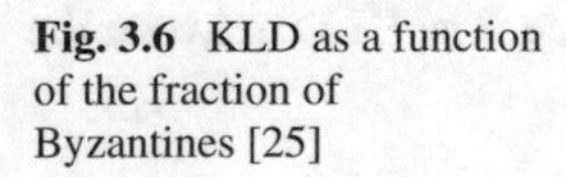
Fig. 3.6 KLD as a function of the fraction of Byzantines [25]

attack strategies. There are two types of Byzantine attacks: independent malicious Byzantine attacks (IMBA) and cooperative malicious Byzantine attacks (CMBA). In an independent attack, each Byzantine attacks the distributed detection system independently relying on its own observation. Since the Byzantines do not know the identity of the other Byzantines in the network, $P_d^B = P_d^H$ and $P_f^B = P_f^H$ which gives $\alpha_{blind} = 0.5$ from Lemma 3.7. This implies that the number of Byzantines needs to be at least 50% to blind the FC when the Byzantines attack the network independently. In CMBA, Byzantines collaborate to make a decision regarding the true hypothesis and use this information to attack the network. Using collaboration, the Byzantines can reduce the minimum critical power α_{blind} by increasing $(P_d^B - P_f^B)$. Assume that the Byzantines are colluding using the "L-out-of M" rule to make their decision; i.e., if more than L-out-of-M Byzantines make the decision "1", then all the collaborating Byzantines in the network send a "0". The value of L is taken to be $\lceil M/2 \rceil$, which corresponds to the majority rule. As Fig. 3.6 shows, α_{blind} decreases with collaboration of the Byzantines.

3.2.3 Analysis from Network Designer's Perspective

Now, let us discuss some countermeasures used to protect the network from these Byzantines. Byzantines can be treated as outliers, and therefore, one may use signal processing techniques to mitigate their effects.

A simple and intuitive method to mitigate the effect of Byzantines is to identify them [25] and then exclude them from decision making. For identification purposes, one needs to observe the behavior of nodes over time. We first discuss some of the schemes proposed in the literature which treat the FC as a watchdog to mitigate the effect of Byzantines.

Reputation-Based Byzantine Mitigation Scheme

A simple and effective scheme to identify the Byzantines is by assigning a reputation score to each node based on the quality of the data they provide [25]. Let us divide the detection process into time windows each consisting of T detection intervals. Next, define a reputation metric κ_i for each node as the number of mismatches in a time window consisting of T sensing periods between ith node's local decision and the global decision made at the FC using the majority rule. The reputation metric is given by

$$\kappa_i = \sum_{t=1}^{T} \mathscr{I}_{(u_i[t] \neq u_0[t])}, \tag{3.32}$$

where $u_i[t]$ is the ith node's local decision at time instant t, $u_0[t]$ is the global decision made at the FC at time instant t, and $\mathscr{I}_{(S)}$ is the indicator function over the set S. The nodes for which this reputation metric κ_i is greater than a pre-determined threshold κ are tagged as Byzantines and removed from the fusion process.

Numerical Results

The effectiveness of the reputation-based mitigation scheme is presented through numerical results. Figure 3.7 shows the isolation of nodes from information fusion at the FC as a function of the number of time windows when $N = 100$, $(P_d, P_f) = (0.8, 0.2)$. Each time window consists of $T = 4$ detection intervals. At $\alpha = 0.4$, in a span of only four time windows, the reputation-based mitigation scheme isolates all the Byzantine attackers without removing a significant number of honest nodes. However when $\alpha = 0.5$, honest and Byzantine nodes are eliminated with equal probability. Therefore, this scheme works only when the number of Byzantines in the network is less than 50% of the total number of nodes since the FC uses the majority rule for fusion. If the Byzantines have a majority in num-

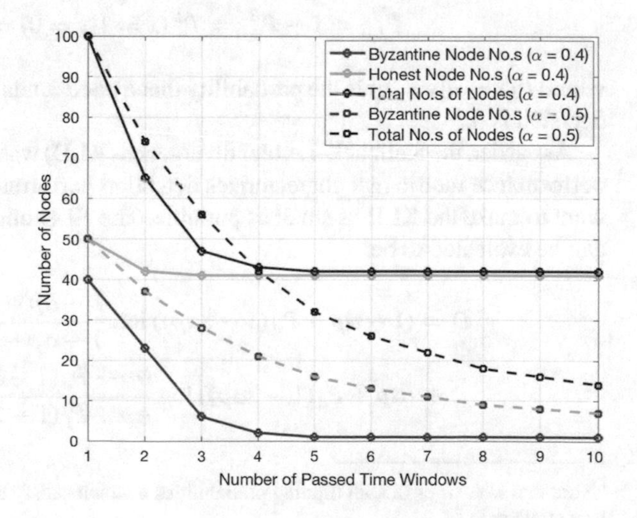

Fig. 3.7 Number of nodes versus number of time windows [25]

ber, the above reputation-based scheme identifies the honest nodes as outliers and removes them from the network and thereby worsens the inference performance of the network.

Other expectation–maximization (EM) algorithm-based schemes have also been proposed in the literature for defense against Byzantine attacks [27, 28]. Noise-enhanced [4] algorithms have been proposed to make a sub-optimal network more robust to Byzantine attacks [7, 34].

3.2.4 Covert Byzantine Attacks

The attacker's only objective in the previous sections was to degrade the detection performance as much as possible. We call such attacks, overt data falsification attacks. It was shown that when the fraction of Byzantines $\alpha \leq 0.5$, flipping their decision with probability $p = 1$ is the best strategy for Byzantines.[6] In other words, an attacker should attack with its full efficacy, $p = 1$, to degrade the network-wide detection performance. Observe that the overt data falsification attackers can be easily detected and discarded because attacking with full efficacy can very easily expose the adversary to the fusion center's defense mechanism. However, knowing the existence of the defense mechanism, a smart adversary can manage to disguise its intention while accomplishing its attack. Such a covert Byzantine attack formulation was considered by the authors in [13]. For tractability, it was assumed that attack strategies $P^H_{j,1}$, $P^H_{j,0}$, and $P^B_{j,1}$, $P^B_{j,0}$ ($j \in \{0, 1\}$) are as follows:

$$P^H_{1,1} = 1 - P^H_{0,1} = P^H(x = 1|y = 1) = 1 \tag{3.33}$$

$$P^H_{1,0} = 1 - P^H_{0,0} = P^H(x = 1|y = 0) = 0 \tag{3.34}$$

$$P^B_{1,1} = 1 - P^B_{0,1} = P^B(x = 1|y = 1) = 1 - p \tag{3.35}$$

$$P^B_{1,0} = 1 - P^B_{0,0} = P^B(x = 1|y = 0) = p\,, \tag{3.36}$$

where $P(x = a|y = b)$ is the probability that a node sends a to the FC when its actual decision is b.

As earlier, the Kullback–Leibler divergence (KLD) was assumed to be the network performance metric that characterizes detection performance. The Byzantine nodes want to make the KLD as small as possible. The KLD under data falsification attack can be evaluated to be:

$$\begin{aligned} D = {} & (1 - \alpha p - P_d(1 - 2\alpha p)) \log \frac{1 - \alpha p - P_d(1 - 2\alpha p)}{1 - \alpha p - P_f(1 - 2\alpha p)} \\ & + (\alpha p + P_d(1 - 2\alpha p)) \log \frac{\alpha p + P_d(1 - 2\alpha p)}{\alpha p + P_f(1 - 2\alpha p)}\,. \end{aligned} \tag{3.37}$$

[6] Note that when $\alpha > 0.5$, all flipping probabilities p which satisfy $\alpha p = 0.5$ will correspond to the best strategy.

Knowing the existence of the defense mechanism, an intelligent adversary would minimize KLD while constraining its exposure to the defense mechanism. The covert data falsification attack problem can be formally stated as

$$\begin{aligned} &\underset{p}{\text{minimize}} \quad D \\ &\text{subject to} \quad P_B^{iso} \leq \gamma \\ &\qquad\qquad\quad 0 \leq p \leq 1 \end{aligned} \tag{P3.4}$$

where P_B^{iso} is the probability that a Byzantine is detected and isolated by the defense mechanism. An attacker tries to maintain P_B^{iso} below a threshold γ to constrain its exposure to the defense/detection scheme. Note that the covert data falsification problem is dependent on the Byzantine detection scheme employed in the system. Hence, the authors in [13] proposed a Byzantine detection scheme and solved the optimization problem as discussed next.

An Efficient Byzantine Detection Scheme

A natural scheme for the mitigation of Byzantine attacks is to identify the Byzantines by observing their behavior over time and decide on whether it behaves closer to an honest or a Byzantine node. Let us assume that the FC observes the local decisions of each node over a time window consisting of T sensing periods, which can be denoted by $U^i = [u_1^i, \ldots, u_T^i]$ for node i. Also, assume that there is one honest anchor node with probability of detection P_d^A and probability of false alarm P_f^A that are known to the FC. The anchor node is employed to provide the gold standard which is used to detect whether or not other nodes are Byzantines. The FC can also serve as an anchor node in that it can directly observe the phenomenon and make a decision. Denote the Hamming distance between reports of the anchor node and node i over the time window of length T by $d_i^A = ||U^A - U^i||$, which represents the number of elements that are different between U^A and U^i. Since the FC is aware of the fact that Byzantines might be present in the network, it compares d_i^A to a threshold η to make the decision regarding the presence of the Byzantines. The authors in [13] counter the data falsification attack by isolating or cutting-off those nodes from the information fusion process whose distance d_i^A is greater than a fixed threshold η. The probability that a Byzantine is isolated at the end of the time window T, P_B^{iso}, is a function of the parameter η, which is under the control of the FC and parameters (p, α), which are under the control of the attacker.

As aforementioned, local decisions of the nodes are compared to the decisions of the anchor node over a time window of length T. The probability that an honest node makes a decision that is different from the anchor node can be derived to be

$$\begin{aligned} &P_{diff}^{AH} \\ &= P(u_i^A = 1, u_i^H = 0, H_0) + P(u_i^A = 0, u_i^H = 1, H_0) \\ &\quad + P(u_i^A = 1, u_i^H = 0, H_1) + P(u_i^A = 0, u_i^H = 1, H_1) \qquad (3.38) \\ &= P_0[(P_f + P_f^A) - 2P_f P_f^A] + P_1[(P_d + P_d^A) - 2P_d P_d^A]\,. \qquad (3.39) \end{aligned}$$

The probability that a Byzantine sends a decision different from that of the anchor node is given by

$$\begin{aligned} &P_{diff}^{AB} \\ &= P(u_i^A = 1, u_i^B = 0, H_0) + P(u_i^A = 0, u_i^B = 1, H_0) \\ &\quad + P(u_i^A = 1, u_i^B = 0, H_1) + P(u_i^A = 0, u_i^B = 1, H_1) \end{aligned} \tag{3.40}$$

$$\begin{aligned} &= P_0[(1-p)\{P_f^A(1-P_f) + (1-P_f^A)P_f\} + p\{P_f^A P_f \\ &\quad + (1-P_f^A)(1-P_f)\}] + P_1[(1-p)\{P_d^A(1-P_d) \\ &\quad + (1-P_d^A)P_d\} + p\{P_d^A P_d + (1-P_d^A)(1-P_d)\}]. \end{aligned} \tag{3.41}$$

The difference between the reports by a sensor and the anchor node is a Bernoulli random variable with mean P_{diff}^{AH} for honest nodes and P_{diff}^{AB} for Byzantines. In this section, we counter the data falsification attack by isolating or cutting-off those nodes from the information fusion process whose distance d_i^A is greater than a fixed threshold η. The probability that a Byzantine is isolated at the end of the time window T can be expressed as

$$P_B^{iso} = P(d_i^A > \eta) = \sum_{j=\eta+1}^{T} \binom{T}{j} (P_{diff}^{AB})^j (1 - P_{diff}^{AB})^{T-j}. \tag{3.42}$$

When T is large enough, by using the normal approximation, we have

$$P_B^{iso} = Q\left(\frac{\eta - TP_{diff}^{AB}}{\sqrt{(TP_{diff}^{AB}(1 - P_{diff}^{AB}))}}\right). \tag{3.43}$$

Similarly, the isolation probability for an honest node can be obtained as

$$P_H^{iso} = Q\left(\frac{\eta - TP_{diff}^{AH}}{\sqrt{(TP_{diff}^{AH}(1 - P_{diff}^{AH}))}}\right). \tag{3.44}$$

The optimal threshold, η, is selected by constraining the isolation probability of honest nodes such that $P_H^{iso} \le \kappa$. That is,

$$\eta^* = Q^{-1}(\kappa)\sqrt{(TP_{diff}^{AH}(1 - P_{diff}^{AH}))} + TP_{diff}^{AH}. \tag{3.45}$$

Optimal Covert Data Falsification Attack

The authors in [13] showed that the KLD under a data falsification attack is a monotonically increasing function of $(1 - 2\alpha p)^2$. Therefore, the covert data falsification attack optimization problem given in (P3.4) is equivalent to

$$\begin{aligned} \underset{p}{\text{minimize}} \quad & (1-2\alpha p)^2 \\ \text{subject to} \quad & P_B^{iso} \leq \gamma \\ & 0 \leq p \leq 1\,. \end{aligned} \tag{P3.5}$$

Solution of the optimization problem (P3.5) can be obtained analytically and is given as follows:

Case 1: When $\alpha \leq 0.5$: If $P_B^{iso}(\alpha, p=1) \leq \gamma$, the covertness constraint is not active and the attacker should flip with probability $p = 1$, similar to the no constraint case. When $P_B^{iso}(\alpha, p=1) > \gamma$, the covertness constraint is active. In this case, the attacker will start decreasing the flipping probability until $P_B^{iso}(\alpha, p=1) = \gamma$. For a given γ and η^* as given in (3.45), the optimal flipping probability p^* is obtained by setting $P_B^{iso} = \gamma$ as given in (3.46).

Case 2: When $\alpha > 0.5$: If $P_B^{iso}(\alpha, p=\frac{1}{2\alpha}) \leq \gamma$, the covertness constraint is not active and the attacker would flip with probability $p = \frac{1}{2\alpha}$, similar to the no constraint case. Otherwise, it would flip with probability p^* given in (3.46).

$$p^* = \frac{M^* - P_0(P_f^A(1-P_f) + (1-P_f^A)P_f) - P_1(P_d^A(1-P_d) + (1-P_d^A)P_d)}{P_0(1-2P_f^A)(1-2P_f) + P_1(1-2P_d^A)(1-2P_d)}, \text{ where } M^* = \max(M(0), M(1)). \tag{3.46}$$

$$M(i) = \frac{T(2\eta^* + (Q^{-1}(\gamma))^2) + (-1)^i\sqrt{T^2(2\eta^* + (Q^{-1}(\gamma))^2)^2 - 4(\eta^*)^2 T(T + (Q^{-1}(\gamma))^2)}}{2(T^2 + (Q^{-1}(\gamma))^2 T)} \text{ for } i = 0, 1 \tag{3.47}$$

Blinding the Fusion Center

Now let us analyze the minimum fraction of Byzantines needed to make $KLD = 0$ or to blind the FC. Note that $KLD = 0$ only when $(1-2\alpha p)^2 = 0$.

Case 1: When $\alpha \leq 0.5$: If $P_B^{iso}(\alpha = 0.5, p=1) \leq \gamma$, the minimum fraction of Byzantines needed to make $KLD = 0$ is 50% or $\alpha_{blind} = 0.5$; otherwise, the attacker cannot blind the FC.

Case 2: When $\alpha > 0.5$: If $P_B^{iso}(\alpha, p=\frac{1}{2\alpha}) \leq \gamma$, the attacker can blind the FC; otherwise, it cannot. In this case, the minimum fraction of Byzantines needed to make $KLD = 0$ is $\alpha_{blind} = \dfrac{1}{2p^*}$, where p^* as given in (3.46).

Numerical Results

The analytical results for covert data falsification attack problem are corroborated numerically. Assume that identical sensors with probability of detection $P_d = 0.6$ and probability of false alarm $P_f = 0.2$ are detecting the presence of the phenomenon. Further, assume that an anchor node with probability of detection $P_d^A = 0.65$ and probability of false alarm $P_f^A = 0.2$ is present in the network. The FC observes local decision of the nodes over a $T = 15$ time window. Reputation metric η^* has been

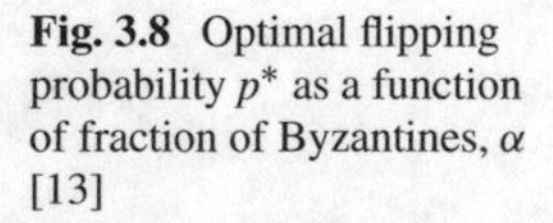

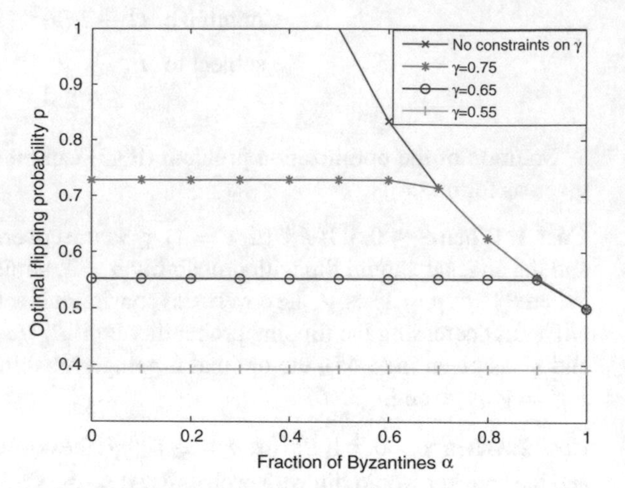

Fig. 3.8 Optimal flipping probability p^* as a function of fraction of Byzantines, α [13]

chosen such that the probability of an honest node being removed from the process at the end of the time window T is $P_H^{iso} < 0.3$, i.e., $\kappa = 0.3$.

Figure 3.8 shows the optimal flipping probability, p^*, as a function of the fraction of Byzantines, α, for different covertness constraints, γ. Observe that, when there is no constraint on γ, flipping with probability $p = 1$ is the best strategy for $\alpha \leq 0.5$ and for $\alpha > 0.5$, all flipping probabilities p which satisfy $\alpha p = 0.5$ are the best strategies. For $\alpha \leq 0.5$, all three covertness constraints considered, $\gamma = [0.55,\ 0.65,\ 0.75]$, are less than $P_B^{iso}(\alpha, p = 1) = 0.8727$ (see Fig. 3.9), and thus, covertness constraints are active. Therefore, the attacker chooses flipping probability p^* given in (3.46) which is less than 1. For $\alpha > 0.5$, the attacker continues to flip with probability p^* until $P_B^{iso}(\alpha, p = \frac{1}{2\alpha}) > \gamma$, and after that point, the attacker flips with probability $\frac{1}{2\alpha}$, similar to the no constraint case. Note that for $\gamma = 0.55$, $P_B^{iso}(\alpha, p = \frac{1}{2\alpha}) \leq \gamma,\ \forall \alpha$, therefore, the optimal flipping probability p^* is the same for all $0 \leq \alpha \leq 1$.

Figure 3.9 shows the probability of a Byzantine node being removed from the fusion process as a function of the fraction of Byzantines in the network. We plot P_B^{iso} for different covertness constraints, γ. Byzantines flip their decision with probability p^* as shown in Fig. 3.8. We observe that the isolation probability of a Byzantine, P_B^{iso}, is a non-increasing function of the fraction of Byzantine attackers α and follows a similar pattern as in Fig. 3.8.

Next Fig. 3.10 shows the effect of the fraction of the Byzantines on detection performance. We plot minimum K-L divergence, $\min_p D$, as a function of the fraction of the Byzantines for different covertness constraints, γ. Observe that in the absence of the covertness constraint, 50% Byzantine nodes are needed to make $D = 0$ or blind the fusion center; however, p_B^{iso} is very high for this case and Byzantine nodes can be easily caught by the FC. For covertness constraints, $\gamma = [0.65,\ 0.75]$, the attacker can still make $D = 0$ or blind the fusion center; however, the fraction of Byzantines needed to blind the FC increases. When covertness constraint $\gamma = 0.55$,

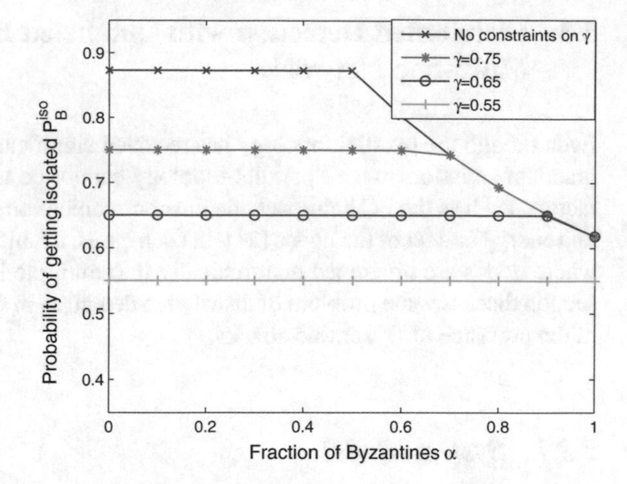

Fig. 3.9 Probability of Byzantine nodes being isolated, P_B^{iso}, as a function of fraction of the Byzantines, α [13]

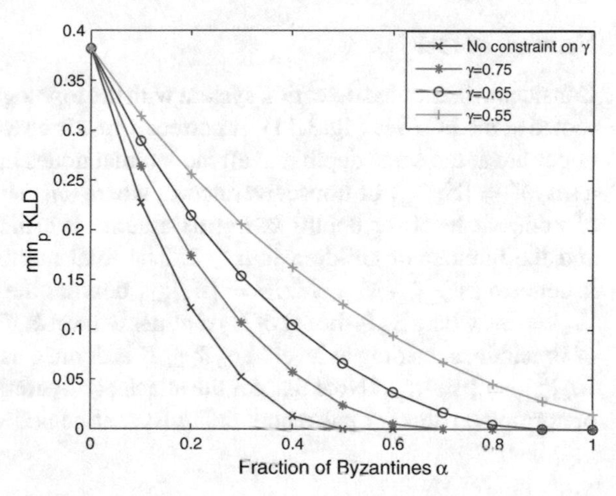

Fig. 3.10 Minimum K-L divergence as a function of the fraction of the Byzantines [13]

the attacker cannot blind the FC; however, it can make KLD very low, $D < 0.025$, and still maintains high covertness.

Numerical results presented in this section suggest that there is a trade-off between attack efficacy and covertness. By attacking with flipping probability p^*, an adversary can degrade the detection performance while constraining its exposure to the defense mechanism.

Other detection frameworks have also been considered in the literature for studying Byzantines in parallel networks, such as game-theoretical formulations in [1, 12], false discovery rate (FDR)-based detection in [35], and audit-bit-based detection in [8, 9].

3.3 Distributed Detection with Unlabeled Byzantine Data: Multi-hop Networks

Even though the parallel topology has received significant attention, there are many practical situations where parallel topology cannot be implemented due to several factors, such as the FC being outside the communication range of the nodes and limited energy budget of the nodes [21]. In such cases, a multi-hop network is employed, where nodes are organized hierarchically into multiple levels (tree networks). This section discusses the problem of distributed detection in multi-hop (or tree) topology in the presence of Byzantine attacks.

3.3.1 System Model

Network Model

Consider a distributed detection system with the topology of a perfect a-tree $T(K, a)$ rooted at the FC (See Fig. 3.11). A perfect a-tree is an a-ary tree in which all the leaf nodes are at the same depth and all the internal nodes have degree "a". $T(K, a)$ has a set $\mathcal{N} = \{\mathbb{N}_k\}_{k=1}^{K}$ of transceiver nodes, where $|\mathbb{N}_k| = N_k = a^k$ is the total number of nodes at level (or depth) k. Let us assume that the depth of the tree is $K > 1$ and the number of children is $a \geq 2$. The total number of nodes in the network is denoted as $\sum_{k=1}^{K} N_k = N$. $\mathscr{B} = \{\mathbb{B}_k\}_{k=1}^{K}$ denotes the set of Byzantine nodes with $|\mathbb{B}_k| = B_k$, where $\mathbb{B}_k$ is the set of Byzantines at level k. The set containing the number of Byzantines residing at levels $1 \leq k \leq K$ is defined as an attack configuration, i.e., $\{B_k\}_{k=1}^{K} = \{|\mathbb{B}_k|\}_{k=1}^{K}$. Note that for the attack configuration $\{B_k\}_{k=1}^{K}$, the total number of corrupted paths (or paths containing Byzantine nodes) from level k to the FC are

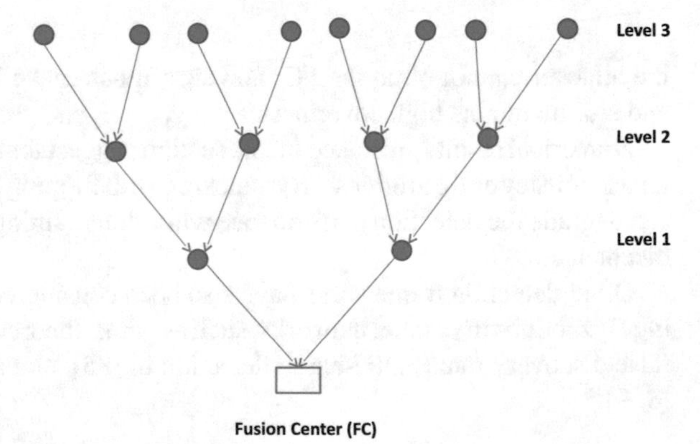

Fig. 3.11 A distributed detection system organized as a perfect binary tree $T(3, 2)$ is shown as an example

$\sum_{i=1}^{k} B_i \frac{N_k}{N_i}$, where $B_i \frac{N_k}{N_i}$ gives the total number of covered[7] nodes at level k by B_i Byzantines at level i. If we denote $\alpha_k = \frac{B_k}{N_k}$, then $\frac{\sum_{i=1}^{k} B_i \frac{N_k}{N_i}}{N_k} = \sum_{i=1}^{k} \alpha_i$ is the fraction of decisions coming from level k that encounter a Byzantine. In practice, nodes operate with very limited energy, and therefore, it is reasonable to assume that the packet IDs (or source IDs) are not forwarded in the tree to save energy. Moreover, even in cases where the packet IDs (or source IDs) are forwarded, note that the packet IDs (or source IDs) can be tampered too, thereby preventing the FC from knowing the exact source of a message. Therefore, assume that the FC looks at messages coming from nodes in a probabilistic manner and considers each received bit to originate from nodes at level k with certain probability $\beta_k \in [0, 1]$. This also implies that, from the FC's perspective, received bits are identically distributed. For a $T(K, a)$,

$$\beta_k = \frac{a^k}{N}. \tag{3.48}$$

Distributed detection in a tree topology

Consider a binary hypothesis testing problem with the two hypotheses H_0 (signal is absent) and H_1 (signal is present). Each node i at level k acts as a source in that it makes a one-bit local decision $v_{k,i} \in \{0, 1\}$ and sends $u_{k,i}$ to its parent node at level $k - 1$, where $u_{k,i} = v_{k,i}$ if i is an uncompromised (honest) node, but for a compromised (Byzantine) node i, $u_{k,i}$ need not be equal to $v_{k,i}$. It also receives the decisions $u_{k',j}$ of all successors j at levels $k' \in [k + 1, K]$, which are forwarded to i by its immediate children. It forwards[8] these received decisions along with $u_{k,i}$ to its parent node at level $k - 1$. If node i is a Byzantine, then it might alter these received decisions before forwarding. It is assumed that the communication channels between children and the parent nodes are error-free. Denote the probabilities of detection and false alarm of a honest node i at level k by $P_d^H = P(v_{k,i} = 1|H_1, i \notin \mathbb{B}_k)$ and $P_{fa}^H = P(v_{k,i} = 1|H_0, i \notin \mathbb{B}_k)$, respectively. Similarly, the probabilities of detection and false alarm of a Byzantine node i at level k are denoted by $P_d^B = P(v_{k,i} = 1|H_1, i \in \mathbb{B}_k)$ and $P_{fa}^B = P(v_{k,i} = 1|H_0, i \in \mathbb{B}_k)$, respectively.

3.3.2 Analysis from Attacker's Perspective

In [14], the authors considered that each Byzantine attacks independently relying on its own observation and derived the fundamental limits as well as designed mitigation strategies for a tree network. First, let us explore the optimal attack strategies for the Byzantines that most degrade the detection performance.

---

[7]Node i at level k' covers all its children at levels $k' + 1 \leq k \leq K$ and the node i itself and, therefore, the total number of covered nodes by $B_{k'}$, Byzantine at level k', is $\frac{B_{k'}}{N_{k'}} \cdot \sum_{i=k'}^{K} N_i$.

[8]IEEE 802.16j mandates tree forwarding and IEEE 802.11s standardizes a tree-based routing protocol. Note that IEEE 802.16j and IEEE 802.11s are standard protocols for tree-based networks.

Attack Model

Now a mathematical model for the Byzantine attack in a tree network is presented. If a node is honest, then it forwards its own decision and received decisions without altering them. However, a Byzantine node, in order to undermine the network performance, may alter its decision as well as received decisions from its children prior to transmission. Let us define the following strategies $P^H_{j,1}$, $P^H_{j,0}$ and $P^B_{j,1}$, $P^B_{j,0}$ ($j \in \{0, 1\}$) for the honest and Byzantine nodes, respectively:

Honest nodes:

$$P^H_{1,1} = 1 - P^H_{0,1} = P^H(x = 1|y = 1) = 1 \tag{3.49}$$

$$P^H_{1,0} = 1 - P^H_{0,0} = P^H(x = 1|y = 0) = 0 \tag{3.50}$$

Byzantine nodes:

$$P^B_{1,1} = 1 - P^B_{0,1} = P^B(x = 1|y = 1) \tag{3.51}$$

$$P^B_{1,0} = 1 - P^B_{0,0} = P^B(x = 1|y = 0) \tag{3.52}$$

where $P(x = a|y = b)$ is the probability that a node sends a to its parent when it receives b from its child or its actual decision is b. Furthermore, assume that if a node (at any level) is a Byzantine then none of its ancestors are Byzantines; otherwise, the effect of a Byzantine due to other Byzantines on the same path may be nullified, e.g., a Byzantine ancestor re-flipping the already flipped decisions of its successor. This means that any path from a leaf node to the FC will have at most one Byzantine. Thus, we have $\sum_{k=1}^{K} \alpha_k \leq 1$ since the average number of Byzantines along any path from a leaf to the root cannot be greater than 1.

Performance metric

The Byzantine attacker always wants to degrade the detection performance at the FC as much as possible; in contrast, the FC wants to maximize the detection performance. The authors in [14] employed the Kullback–Leibler divergence (KLD) to be the network performance metric that characterizes detection performance. The KLD between the distributions $\pi_{j,0} = P(z = j|H_0)$ and $\pi_{j,1} = P(z = j|H_1)$ can be expressed as

$$D(\pi_{j,1}||\pi_{j,0}) = \sum_{j\in\{0,1\}} P(z = j|H_1) \log \frac{P(z = j|H_1)}{P(z = j|H_0)}. \tag{3.53}$$

$$\begin{aligned} P(z_i = j|H_0) = &\left[\sum_{k=1}^{K} \beta_k \left(\sum_{i=1}^{k} \alpha_i\right)\right] [P^B_{j,0}(1 - P^B_{fa}) + P^B_{j,1} P^B_{fa}] \\ &+ \left[\sum_{k=1}^{K} \beta_k \left(1 - \sum_{i=1}^{k} \alpha_i\right)\right] [P^H_{j,0}(1 - P^H_{fa}) + P^H_{j,1} P^H_{fa}] \end{aligned} \tag{3.54}$$

$$P(z_i = j|H_1) = \left[\sum_{k=1}^{K} \beta_k \left(\sum_{i=1}^{k} \alpha_i\right)\right] [P_{j,0}^B(1 - P_d^B) + P_{j,1}^B P_d^B]$$
$$+ \left[\sum_{k=1}^{K} \beta_k \left(1 - \sum_{i=1}^{k} \alpha_i\right)\right] [P_{j,0}^H(1 - P_d^H) + P_{j,1}^H P_d^H] \quad (3.55)$$

For a K-level network, distributions of received decisions at the FC $z_i, i = 1, .., N$, under H_0 and H_1 are given by (3.54) and (3.55), respectively.

In order to make the analysis tractable, assume that the network designer attempts to maximize the KLD of each node as seen by the FC. On the other hand, the attacker attempts to minimize the KLD of each node as seen by the FC.

Blinding the Fusion Center

As discussed earlier, the Byzantine nodes attempt to make their KL divergence as small as possible. Since the KLD is always nonnegative, Byzantines attempt to choose $P(z = j|H_0)$ and $P(z = j|H_1)$ such that KLD is zero. In this case, an adversary can falsify the data that the FC receives from the nodes such that no information is conveyed. In [14], the authors derived an analytical condition to blind the FC for a tree network of an arbitrary depth.

Proposition 3.1 ([14]) *In a tree network with K levels, there exists an attack probability distribution* $(P_{0,1}^B, P_{1,0}^B)$ *that can make* $KLD = 0$*, and thereby blind the FC, if and only if* $\{B_k\}_{k=1}^K$ *satisfy*

$$\sum_{k=1}^{K} \left(\frac{B_k}{N_k} \sum_{i=k}^{K} N_i\right) \geq \frac{N}{2}. \quad (3.56)$$

Proof The proof is given in Appendix A.5. □

Dividing both sides of (3.56) by N, the above condition can be written as $\sum_{k=1}^{K} \beta_k \sum_{i=1}^{k} \alpha_i \geq 0.5$. This implies that to make the FC blind, 50% or more nodes in the network need to be covered by the Byzantines. Observe that Proposition 3.1 suggests that there exist multiple attack configurations $\{B_k\}_{k=1}^K$ that can blind the FC. Also note that some of these attack sets require Byzantines to compromise less than 50% of the nodes in the network. For example, attacking half of the nodes at level 1, i.e., $B_1 = \frac{N_1}{2} << \frac{N}{2}$, cover 50% of the nodes in the network, and therefore, the FC becomes blind. This implies that in the tree topology, Byzantines have more degrees of freedom to blind the FC as compared to the parallel topology.

Optimal Attack Strategies

The authors in [14] further explored the optimal attack probability distribution $(P_{0,1}^B, P_{1,0}^B)$ that minimizes KLD when the FC cannot be made blind. For analytical tractability, assume $P_d^H = P_d^B = P_d$ and $P_{fa}^H = P_{fa}^B = P_{fa}$. It was shown that attacking with symmetric flipping probabilities is the optimal strategy in the region where the attacker cannot blind the FC. In other words, attacking with $P_{1,0} = P_{0,1}$ is the optimal strategy for the Byzantines.

Lemma 3.9 ([14]) *In the region where the attacker cannot blind the FC, the optimal attack strategy comprises symmetric flipping probabilities. More specifically, any nonzero deviation* $\varepsilon_i \in (0, p]$ *in flipping probabilities* $(P_{0,1}^B, P_{1,0}^B) = (p - \varepsilon_1, p - \varepsilon_2)$, *where* $\varepsilon_1 \neq \varepsilon_2$, *will result in an increase in the KLD.*

Proof The proof is given in Appendix A.6. □

The authors in [14] also derived a closed-form expression for the optimal attack probability distribution $(P_{j,1}^B, P_{j,0}^B)$ that minimizes *KLD* in the region where the attacker cannot blind the FC.

Theorem 3.4 ([14]) *In the region where the attacker cannot blind the FC, the optimal attack strategy is given by* $(P_{0,1}^B, P_{1,0}^B) = (1, 1)$.

Proof The proof is given in Appendix A.7. □

Numerical Results

Next, to gain insights into the solution, some numerical results are presented. Figure 3.12 shows KLD as a function of the flipping probabilities $(P_{1,0}^B, P_{0,1}^B)$. It is assumed that the probability of detection is $P_d = 0.8$, the probability of false alarm is $P_{fa} = 0.2$ and the fraction of covered nodes $(t = \sum_{k=1}^{K} [\beta_k (\sum_{i=1}^{k} \alpha_i)])$ by the Byzantines is $t = 0.4$. It can be seen that the optimal attack strategy comprises symmetric flipping probabilities and is given by $(P_{0,1}^B, P_{1,0}^B) = (1, 1)$, which corroborates the theoretical result presented in Lemma 3.9 and Theorem 3.4.

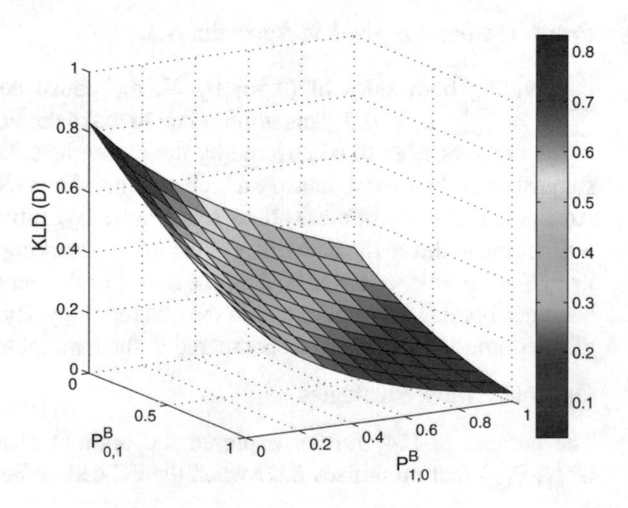

Fig. 3.12 KL distance versus flipping probabilities when $P_d = 0.8$, $P_{fa} = 0.2$, and the fraction of covered nodes by the Byzantines is $t = 0.4$ [14]

3.3.3 Analysis from Network Designer's Perspective

Until now, the problem from the attacker's perspective has been discussed. Now let us look into the problem from a network designer's perspective and discuss techniques to mitigate the effect of the Byzantines.

Robust Topology Design

The authors in [14] assumed that there is a cost associated with attacking each node in the tree (which may represent resources required for capturing a node or cloning a node in some cases) and formulated a robust tree design problem under a cost budget. Let c_k be the cost of attacking any one node at level k and $c_k > c_{k+1}$ for $k = 1, \cdots, K-1$; i.e., it is more costly to attack nodes that are closer to the FC. Observe that a node i at level k covers (in other words, can alter the decisions of) all its successors and node i itself. Also, assume that the network designer has a cost budget $C_{budget}^{network}$ and the attacker has a cost budget $C_{budget}^{attacker}$. The FC designs the network, such that, given the attacker's budget, the fraction of covered nodes is minimized, and consequently, a more robust perfect a-ary tree in terms of KLD[9] is generated.

Since the attacker aims to maximize the fraction of covered nodes by attacking/capturing $\{B_k\}_{k=1}^K$ nodes within the cost budget $C_{budget}^{attacker}$, the FC's objective is to minimize the fraction of covered nodes by choosing the parameters (K, a) optimally in a perfect a-ary tree topology $T(K, a)$ under its cost budget $C_{budget}^{network}$. This situation can be interpreted as a bi-level optimization problem, where the first decision maker (the so-called leader) has the first choice, and the second one (the so-called follower) reacts optimally to the leader's selection. It is the leader's aim to find such a decision which, together with the optimal response of the follower, optimizes the objective function of the leader. Let us assume that the FC has complete information about the attacker's problem, i.e., the objective function and the constraints. Similarly, the attacker is assumed to be aware about the FC's resources, i.e., cost of deploying the nodes $\{c_k\}_{k=1}^K$. Next, robust perfect a-ary tree topology problem can be formalized as follows:

$$
\begin{aligned}
\underset{(K,\,a)\in\mathbb{Z}^+}{\text{minimize}}\ & \frac{\sum_{k=1}^{K}(a^{K-k+1}-1)B_k}{a(a^K-1)} \\
\text{subject to}\ & a_{min} \le a \le a_{max} \\
& K \ge K_{min} \\
& \textstyle\sum_{k=1}^{K} a^k \ge N_{min} \\
& \textstyle\sum_{k=1}^{K} c_k a^k \le C_{budget}^{network} \\
& \underset{B_k\in\mathbb{Z}^+}{\text{maximize}}\ \frac{\sum_{k=1}^{K}(a^{K-k+1}-1)B_k}{a(a^K-1)} \\
& \text{subject to}\ \textstyle\sum_{k=1}^{K} c_k B_k \le C_{budget}^{attacker} \\
& \qquad B_k \le a^k, \forall k = 1, 2, \ldots, K
\end{aligned}
\tag{P3.6}
$$

[9]It was shown in [14] that by minimizing/maximizing the fraction of covered nodes, the FC can maximize/minimize the KLD. Using this fact, the authors considered fraction of covered nodes in lieu of the KLD in their analysis.

where $\mathbb{Z}^+$ is the set of nonnegative integers, $a_{min} \geq 2$ and $K_{min} \geq 2$. The objective function for the FC is the fraction of covered nodes by the Byzantines $\frac{\sum_{k=1}^{K} P_k B_k}{\sum_{k=1}^{K} N_k}$, where $P_k = \frac{a^{K-k+1}-1}{a-1}$ and $\sum_{k=1}^{K} N_k = \frac{a(a^K-1)}{a-1}$. In the constraint $a_{min} \leq a \leq a_{max}$, a_{max} represents the hardware constraint imposed by the Medium Access Control (MAC) scheme used and a_{min} represents the design constraint enforced by the FC. The constraint on the number of nodes in the network $\sum_{k=1}^{K} a^k \geq N_{min}$ ensures that the network satisfies pre-specified detection performance guarantees. In other words, N_{min} is the minimum number of nodes needed to guarantee a certain detection performance. The constraint on the cost expenditure $\sum_{k=1}^{K} c_k a^k \leq C_{budget}^{network}$ ensures that the total expenditure of the network designer does not exceed the available budget.

For an attacker, the objective function is the same as that of the FC, but the sense of optimization is opposite, i.e., maximization of the fraction of covered nodes. The constraint $\sum_{k=1}^{K} c_k B_k \leq C_{budget}^{attacker}$ ensures that the total expenditure of the attacker does not exceed the available budget. The constraints $B_k \leq a^k$, $\forall k$ are logical conditions, which prevent the attacker from attacking non-existing resources.

Note that the bi-level optimization problem, in general, is an NP-hard problem [2]. In fact, the optimization problem corresponding to the attacker is the packing formulation of the Bounded Knapsack Problem (BKP) [5], which itself, in general, is NP-hard. The authors in [14] derived some properties of the objective function that enabled robust topology design problem to have a polynomial time solution.

Lemma 3.10 ([14]) *In a perfect a-ary tree topology, the fraction of covered nodes* $\frac{\sum_{k=1}^{K} P_k B_k}{\sum_{k=1}^{K} N_k}$ *by the attacker with the cost budget* $C_{budget}^{attacker}$ *for an optimal attack is a non-decreasing function of the number of levels K in the tree.*

Proof The proof is given in Appendix A.8. □

Next, let us explore some properties of the fraction of covered nodes with parameter a for a *perfect a-ary* tree topology. Let us define a parameter a_{min} as follows. For a fixed K and attacker's cost budget $C_{budget}^{attacker}$, a_{min} is defined as the minimum value of a for which the attacker cannot blind the network or cover 50% or more nodes. Now, the analysis can be restricted to $a_{min} \leq a \leq a_{max}$. Note that the attacker cannot blind all the trees $T(K, a)$ for which $a \geq a_{min}$ and can blind all the trees $T(K, a)$ for which $a < a_{min}$.

Lemma 3.11 ([14]) *In a perfect a-ary tree topology, the fraction of covered nodes* $\frac{\sum_{k=1}^{K} P_k B_k}{\sum_{k=1}^{K} N_k}$ *by an attacker with cost budget* $C_{budget}^{attacker}$ *in an optimal attack is a decreasing function of parameter a for a perfect a-ary tree topology for* $a \geq a_{min} \geq 2$.

Proof The proof is given in Appendix A.9. □

Numerical Results

To gain insights into the solution, Figs. 3.13 and 3.14 present some numerical results. Figure 3.13 shows the fraction of covered nodes by the Byzantines as a function of

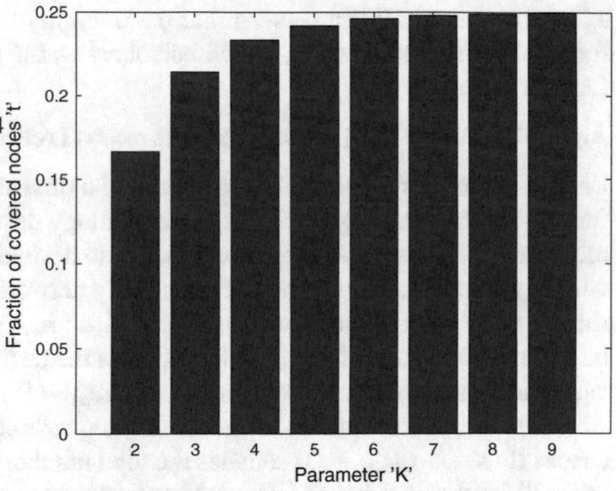

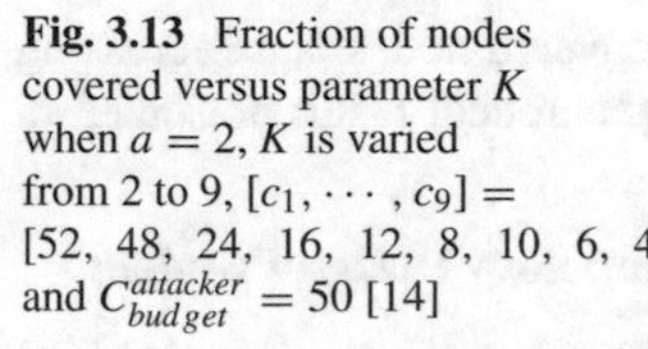

Fig. 3.13 Fraction of nodes covered versus parameter K when $a = 2$, K is varied from 2 to 9, $[c_1, \cdots, c_9] =$ $[52,\ 48,\ 24,\ 16,\ 12,\ 8,\ 10,\ 6,\ 4]$ and $C^{attacker}_{budget} = 50$ [14]

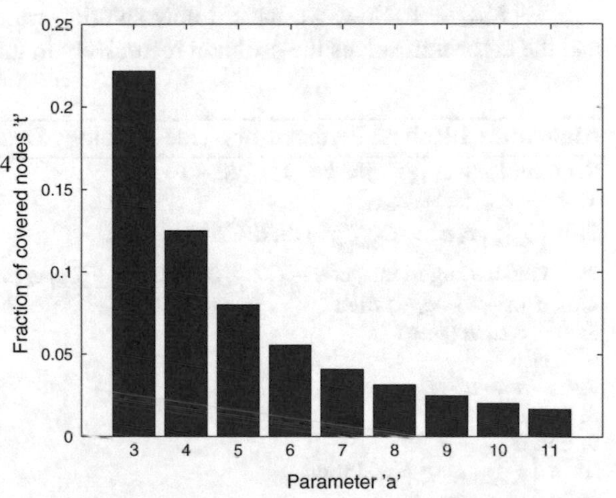

Fig. 3.14 Fraction of nodes covered versus parameter a when $K = 6$, parameter a is varied from 3 to 11, $[c_1, \cdots, c_9] =$ $[52,\ 48,\ 24,\ 16,\ 12,\ 8,\ 10,\ 6,\ 4]$ and $C^{attacker}_{budget} = 50$ [14]

the total number of levels in the tree. Assume that $a = 2$ and vary K from 2 to 9. Also assume that the cost to attack nodes at different levels is given by $[c_1, \cdots, c_9] = [52,\ 48,\ 24,\ 16,\ 12,\ 8,\ 10,\ 6,\ 4]$ and the cost budget of the attacker is $C^{attacker}_{budget} = 50$. For each $T(K,\ 2)$, the optimal attack configuration $\{B_k\}_{k=1}^{K}$ can be found by an exhaustive search. It can be seen that the fraction of covered nodes is a non-decreasing function of the number of levels K, which corroborates the theoretical result presented in Lemma 3.10.

Figure 3.14 shows the fraction of covered nodes by the Byzantines as a function of the parameter a in the tree. Assume that the parameter $K = 6$ and vary a from 3 to 11. Also assume that the cost to attack nodes at different levels is given by $[c_1, \cdots, c_9] = [52,\ 48,\ 24,\ 16,\ 12,\ 8,\ 10,\ 6,\ 4]$ and the cost budget of the attacker is $C^{attacker}_{budget} = 50$. For each $T(6,\ a)$, the optimal attack configuration $\{B_k\}_{k=1}^{K}$ was found by an exhaustive search. It can be seen that the fraction of covered nodes is a decreasing function of the parameter a, which corroborates the theoretical result presented in Lemma 3.11.

Algorithm for Solving Robust Perfect a-ary Tree Topology Design Problem

Based on Lemmas 3.10 and 3.11, a polynomial time algorithm was presented in [14] for solving the robust perfect a-ary tree topology design problem. Observe that the robust network design problem is equivalent to designing perfect a-ary tree topology with minimum K and maximum a that satisfy network designer's constraints. Algorithm 1 starts with the solution candidate $(K_{min},\ a_{max})$. First, the algorithm finds the largest integer $(a_{max} - l),\ l \geq 0$ that satisfies the cost expenditure constraint. If this value violates the hardware constraint, i.e., $(a_{max} - l) < a_{min}$, it will not have any feasible solution which satisfies the network designer's constraints. Next, the algorithm checks if $(K_{min},\ (a_{max} - l))$ satisfies the total number of nodes constraint. If it does, this will be the solution for the problem; otherwise, we increase K_{min} by one, i.e., $K_{min} \leftarrow K_{min} + 1$. Now, we have a new solution candidate $(K_{min} + 1,\ (a_{max} - l))$ and the algorithm solves the problem recursively in this manner.

Algorithm 1 Robust Perfect a-ary Tree Topology Design

Require: $c_k > c_{k+1} \quad for\ k = 1, ..., K-1$
1: $K \leftarrow K_{min}$; $a \leftarrow a_{max}$
2: **if** $\left(\sum_{k=1}^{K} c_k a^k > C_{budget}^{network}\right)$ **then**
3: Find the largest integer $a - \ell, \ell \geq 0$, such that $\sum_{k=1}^{K} c_k (a-\ell)^k \leq C_{budget}^{network}$
4: **if** $(a - \ell < a_{min})$ **then**
5: **return** (ϕ, ϕ)
6: **else**
7: $a \leftarrow a - \ell$
8: **end if**
9: **end if**
10: **if** $\left(\sum_{k=1}^{K} a^k \geq N_{min}\right)$ **then**
11: **return** (K, a)
12: **else**
13: $K \leftarrow K + 1$
14: **return to** Step 2
15: **end if**

This procedure greatly reduces the complexity because we do not need to solve the lower-level problem in this case. Provable optimality of Algorithm 1 was also established in [14].

Numerical Results

The effectiveness of the algorithm is also established through numerical results. Figure 3.15 shows the $\min_{P_{1,0}, P_{0,1}}$ KLD for all the combinations of parameter K and a in the tree. The parameter K was varied from 2 to 10 and a was varied from 3 to 11. Also assume that the costs to attack nodes at different levels are given

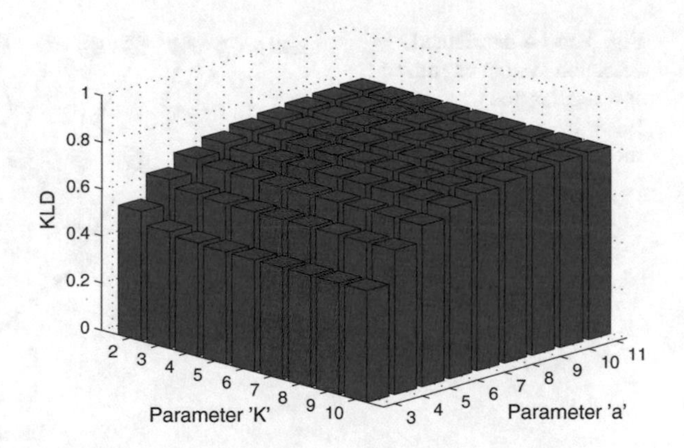

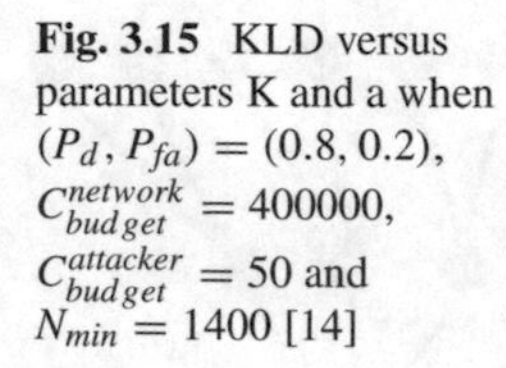

Fig. 3.15 KLD versus parameters K and a when $(P_d, P_{fa}) = (0.8, 0.2)$, $C_{budget}^{network} = 400000$, $C_{budget}^{attacker} = 50$ and $N_{min} = 1400$ [14]

by $[c_1, \cdots, c_{10}] = [52,\ 50,\ 25,\ 24,\ 16,\ 10,\ 8,\ 6,\ 5,\ 4]$, and cost budgets of the network and the attacker are given by $C_{budget}^{network} = 400000$, $C_{budget}^{attacker} = 50$, respectively. The node budget constraint is assumed to be $N_{min} = 1400$. For each $T(K,\ a)$, the optimal attack configuration $\{B_k\}_{k=1}^{K}$ can be found by an exhaustive search. All the feasible solutions are plotted in red and unfeasible solutions are plotted in blue. Note that $T(K_{min},\ a_{max})$ which is $T(2,\ 11)$ is not a feasible solution, and therefore, if we use Algorithm 1, it will try to find the feasible solution which has minimum possible deviation from $T(K_{min},\ a_{max})$. It can be seen that the optimal solution $T(3,\ 11)$ has minimum possible deviation from the $T(K_{min},\ a_{max})$, which corroborates the efficacy of the algorithm.

3.4 Distributed Detection with Labeled Byzantine Data: Multi-hop Networks

In contrast to the previous section, in this section, the problem of distributed detection in regular tree networks[10] with Byzantines is addressed in a practical setup where the FC has the knowledge of which bit is transmitted from which node. Note that in practice, the FC can obtain this information by using MAC schemes[11] and can utilize this information to improve system performance.

[10]For a regular tree, intermediate nodes at different levels are allowed to have different degrees, i.e., number of children.

[11]In practice, one possible way to achieve this is by using the buffer-less TDMA MAC protocol, in which, distinct non-overlapping time slots are assigned (scheduled) to the nodes for communication. One practical example of such a scheme is given in [29].

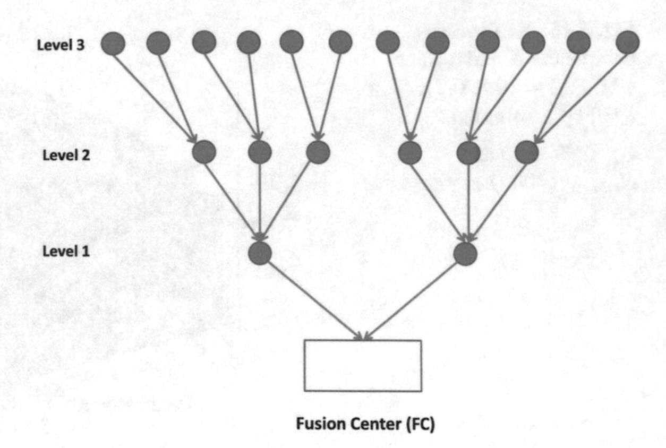

Fig. 3.16 A distributed detection system organized as a regular tree ($a_1 = 2,\ a_2 = 3,\ a_3 = 2$) is shown as an example

3.4.1 System Model

Network Model

Consider a distributed detection system organized as a regular tree network rooted at the FC (see Fig. 3.16). For a regular tree, all the leaf nodes are at the same level (or depth) and all the intermediate nodes at level k have degree a_k. The regular tree is assumed to have a set $\mathscr{N} = \{\mathbb{N}_k\}_{k=1}^{K}$ of transceiver nodes, where $|\mathbb{N}_k| = N_k$ is the total number of nodes at level k. Assume that the depth of the tree is $K > 1$ and $a_k \geq 2$. The total number of nodes in the network is denoted as $N = \sum_{k=1}^{K} N_k$ and $\mathscr{B} = \{\mathbb{B}_k\}_{k=1}^{K}$ denotes the set of Byzantine nodes with $|\mathbb{B}_k| = B_k$, where $\mathbb{B}_k$ is the set of Byzantines at level k. The set containing the number of Byzantines residing at each level k, $1 \leq k \leq K$, is referred to as an attack configuration, i.e., $\{B_k\}_{k=1}^{K} = \{|\mathbb{B}_k|\}_{k=1}^{K}$. Next, the *modus operandi* of the nodes is defined.

Modus Operandi of the Nodes

Consider a binary hypothesis testing problem with two hypotheses H_0 (signal is absent) and H_1 (signal is present). Under each hypothesis, it is assumed that the observations $Y_{k,i}$ at each node i at level k are conditionally independent. Each node i at level k acts as a source in the sense that it makes a one-bit (binary) local decision $v_{k,i} \in \{0, 1\}$ regarding the absence or presence of the signal using the likelihood-ratio test (LRT)[12]

$$\frac{p_{Y_{k,i}}^{(1)}(y_{k,i})}{p_{Y_{k,i}}^{(0)}(y_{k,i})} \underset{v_{k,i}=0}{\overset{v_{k,i}=1}{\gtrless}} \lambda_k, \tag{3.57}$$

where λ_k is the threshold used at level k (it is assumed that all the nodes at level k use the same threshold λ_k) and $p_{Y_{k,i}}^{(j)}(y_{k,i})$ is the conditional probability density function (PDF) of observation $y_{k,i}$ under hypothesis H_j for $j \in \{0, 1\}$. Let us denote the

[12] Note that under the conditional independence assumption, the optimal decision rule at the local sensor is a likelihood-ratio test [32].

probabilities of detection and false alarm of a node at level k by $P_d^k = P(v_{k,i} = 1|H_1)$ and $P_{fa}^k = P(v_{k,i} = 1|H_0)$, respectively, which are functions of λ_k and hold for both Byzantines and honest nodes. After making its one-bit local decision $v_{k,i} \in \{0, 1\}$, node i at level k sends $u_{k,i}$ to its parent node at level $k - 1$, where $u_{k,i} = v_{k,i}$ if i is an honest node, but for a Byzantine node i, $u_{k,i}$ need not be equal to $v_{k,i}$. Node i at level k also receives the decisions $u_{k',j}$ of all successors j at levels $k' \in [k + 1, K]$, which are forwarded to node i by its immediate children, and forwards them to its parent node at level $k - 1$.

Binary Hypothesis Testing at the Fusion Center

Consider the distributed detection problem under the Neyman–Pearson (NP) criterion. The FC receives the decision vectors, $[\mathbf{z_1}, \cdots, \mathbf{z_K}]$, where $\mathbf{z_k}$ for $k \in \{1, \cdots, K\}$ is a decision vector with its elements being $z_1, \cdots, z_{N_k}$, from the nodes at different levels of the tree. Then the FC makes the global decision about the phenomenon by employing the LRT. Due to system vulnerabilities, some of the nodes may be captured by the attacker and reprogrammed to transmit false information to the FC to degrade detection performance. Assume that the only information available at the FC is the probability $\beta_{\bar{x},x}^k$, which is the probability with which the data coming from level k has been falsified. Using this information, the FC calculates the probabilities $\pi_{j,0}^k = P(z_i = j|H_0, k)$ and $\pi_{j,1}^k = P(z_i = j|H_1, k)$, which are the distributions of received decisions z_i originating from level k and arriving at the FC under hypotheses H_0 and H_1. The FC makes its decision regarding the absence or presence of the signal using the following likelihood-ratio test

$$\prod_{k=1}^{K} \left(\frac{\pi_{1,1}^k}{\pi_{1,0}^k} \right)^{s_k} \left(\frac{1 - \pi_{1,1}^k}{1 - \pi_{1,0}^k} \right)^{N_k - s_k} \underset{H_0}{\overset{H_1}{\gtrless}} \eta \tag{3.58}$$

where s_k is the number of decisions that are equal to one and originated from level k, and the threshold η is chosen in order to minimize the missed detection probability (P_M) while keeping the false alarm probability (P_F) below a fixed value δ.

The authors in [15] derived a closed-form expression for the optimal missed detection error exponent for tree networks in the presence of Byzantines and used it as a surrogate for the probability of missed detection.

Proposition 3.2 ([15]) *For a K-level tree network employing the detection scheme as given in (3.58), the asymptotic detection performance, i.e., $N_1 \to \infty$, can be characterized using the missed detection error exponent given below*

$$D = \sum_{k=1}^{K} N_k \left[\sum_{j \in \{0,1\}} \pi_{j,0}^k \log \frac{\pi_{j,0}^k}{\pi_{j,1}^k} \right]. \tag{3.59}$$

Note that (3.59) can be compactly written as $\sum_{k=1}^{K} N_k D_k(\pi_{j,1}^k || \pi_{j,0}^k)$ with $D_k(\pi_{j,1}^k || \pi_{j,0}^k)$ being the KLD between the data coming from node i at level k under

H_0 and H_1. Now, the optimal attack strategies for the Byzantines that degrade the detection performance most can be explored by minimizing the KLD.

3.4.2 Analysis from Attacker's Perspective

In practice, the FC wants to maximize the detection performance, while the Byzantine attacker wants to degrade the detection performance as much as possible which can be achieved by maximizing and minimizing the KLD, respectively. In [15], the authors derived the fundamental limits of distributed detection in tree networks with labeled data in the presence of Byzantines and designed mitigation strategies.

Attack Model

Now a mathematical model for the Byzantine attack in a tree network is presented. If a node is honest, then it forwards its own decision and received decisions without altering them. However, a Byzantine node, in order to undermine the network performance, may alter its decision as well as received decisions from its children prior to transmission. Let us define the following strategies $P^H_{j,1}(k)$, $P^H_{j,0}(k)$ and $P^B_{j,1}(k)$, $P^B_{j,0}(k)$ ($j \in \{0, 1\}$ and $k = 1, \cdots, K$) for the honest and Byzantine nodes at level k, respectively:

Honest nodes:

$$P^H_{1,1}(k) = 1 - P^H_{0,1}(k) = P^H_k(x = 1|y = 1) = 1 \tag{3.60}$$

$$P^H_{1,0}(k) = 1 - P^H_{0,0}(k) = P^H_k(x = 1|y = 0) = 0 \tag{3.61}$$

Byzantine nodes:

$$P^B_{1,1}(k) = 1 - P^B_{0,1}(k) = P^B_k(x = 1|y = 1) \tag{3.62}$$

$$P^B_{1,0}(k) = 1 - P^B_{0,0}(k) = P^B_k(x = 1|y = 0) \tag{3.63}$$

where $P_k(x = a|y = b)$ is the conditional probability that a node at level k sends a to its parent when it receives b from its child or its actual decision is b. For notational convenience, we use $(P^k_{1,0}, P^k_{0,1})$ to denote the flipping probability of the Byzantine node at level k. Furthermore, assume that if a node (at any level) is a Byzantine, then none of its ancestors and successors are Byzantine (non-overlapping attack configuration); otherwise, the effect of a Byzantine due to other Byzantines on the same path may be nullified, e.g., Byzantine ancestor may re-flip the already flipped decisions of its successors. This means that every path from a leaf node to the FC will have at most one Byzantine. Note that for the attack configuration $\{B_k\}_{k=1}^K$, the total number of corrupted paths, i.e., paths containing a Byzantine node, from level k to the

FC are $\sum_{i=1}^{k} B_i \frac{N_k}{N_i}$, where $B_i \frac{N_k}{N_i}$ is the total number of nodes covered[13] at level k by the presence of B_i Byzantines at level i. If we denote $\alpha_k = \frac{B_k}{N_k}$, then $\frac{\sum_{i=1}^{k} B_i \frac{N_k}{N_i}}{N_k} = \sum_{i=1}^{k} \alpha_i$ is the fraction of decisions coming from level k that encounter a Byzantine along the way to the FC. For a large network, due to the law of large numbers, one can approximate the probability that the FC receives the flipped decision $\bar{x}$ of a given node at level k when its actual decision is x as $\beta_{\bar{x},x}^{k} = \sum_{j=1}^{k} \alpha_j P_{\bar{x},x}^{j}$, $x \in \{0, 1\}$.

Blinding the Fusion Center

The authors in [15] derived an analytical condition to blind the FC.

Lemma 3.12 ([15]) *In a tree network with K levels, the minimum number of Byzantines needed to make the Kullback–Leibler divergence (KLD) between the distributions $P(\mathbf{Z}|H_0)$ and $P(\mathbf{Z}|H_1)$ equal to zero (or to make $D_k = 0$, $\forall k$) is given by $B_1 = \lceil \frac{N_1}{2} \rceil$.*

Proof The proof is given in Appendix A.10. □

The authors in [15] further explored the optimal attack probability distribution $(P_{0,1}^{k}, P_{1,0}^{k})$ that minimizes D_k when $\sum_{j=1}^{k} \alpha_j < 0.5$, i.e., in the case where the attacker cannot make $D = 0$.

Optimal Byzantine Attack

Let us first investigate the properties of D_k with respect to $(P_{0,1}^{k}, P_{1,0}^{k})$ assuming $(P_{0,1}^{j}, P_{1,0}^{j})$, $1 \leq j \leq k-1$ to be fixed. It will be shown that attacking with symmetric flipping probabilities is the optimal strategy in the region where the attacker cannot make $D_k = 0$. In other words, attacking with $P_{1,0}^{k} = P_{0,1}^{k}$ is the optimal strategy for the Byzantines.

Lemma 3.13 ([15]) *In the region where the attacker cannot make $D_k = 0$, i.e., for $\sum_{j=1}^{k} \alpha_j < 0.5$, the optimal attack strategy comprises symmetric flipping probabilities $(P_{0,1}^{k} = P_{1,0}^{k} = p)$. In other words, any nonzero deviation $\varepsilon_i \in (0, p]$ in flipping probabilities $(P_{0,1}^{k}, P_{1,0}^{k}) = (p - \varepsilon_1, p - \varepsilon_2)$, where $\varepsilon_1 \neq \varepsilon_2$, will result in an increase in D_k.*

Using the result in Lemma 3.13, the solution for the optimal attack probability distribution $(P_{j,1}^{k}, P_{j,0}^{k})$ that minimizes D_k can be obtained.

Theorem 3.5 ([15]) *In the region where the attacker cannot make $D_k = 0$, i.e., for $\sum_{j=1}^{k} \alpha_j < 0.5$, the optimal attack strategy is given by $(P_{0,1}^{k}, P_{1,0}^{k}) = (1, 1)$.*

Proof The proof is given in Appendix A.11. □

---

[13]Node i at level k' covers (or can alter the decisions of) all its children at levels $k' + 1$ to K and itself. In other words, the total number of covered nodes is equivalent to the total number of corrupted paths (i.e., paths containing a Byzantine node) in the network.

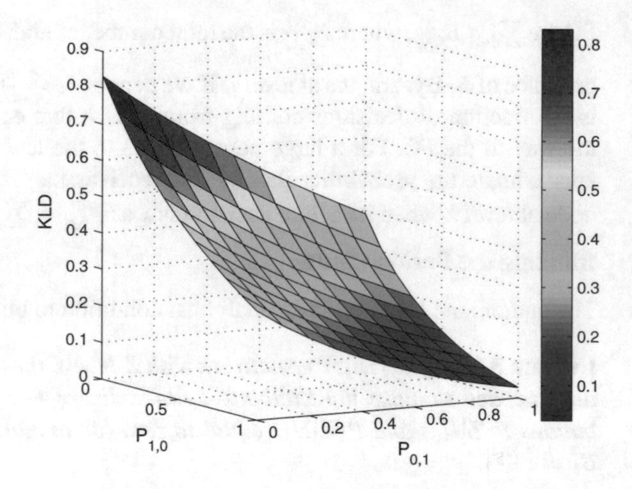

Fig. 3.17 KLD D_k versus flipping probabilities when $P_d^k = 0.8, P_{fa}^k = 0.2$, and the probability that the bit coming from level k encounters a Byzantine is $\sum_{j=1}^{k} \alpha_j = 0.4$ [15]

Thus, for all k,

$$D_k(P_{0,1}^k, P_{1,0}^k) \geq D_k(1, 1). \tag{3.64}$$

Now, by multiplying both sides of (3.64) by N_k and summing it over all K, it can be shown that the KLD, D, is minimized by $(P_{0,1}^k, P_{1,0}^k) = (1, 1)$, for all k, in the region $\sum_{k=1}^{K} \alpha_k < 0.5$.

Furthermore, it was shown in [15] that minimization/maximization of $\sum_{j=1}^{k} \alpha_j$ is equivalent to minimization/maximization of D_k. Using this fact, one can consider the probability that the bit coming from level k encounters a Byzantine (i.e., $t = \sum_{j=1}^{k} \alpha_j$) in lieu of D_k for optimizing the system performance.

Numerical Results

Next, to gain insights into the solution, Fig. 3.17 shows D_k as a function of the flipping probabilities $(P_{1,0}^k, P_{0,1}^k)$. Assume that the probability of detection is $P_d^k = 0.8$, the probability of false alarm is $P_{fa}^k = 0.2$, and the probability that the bit coming from level k encounters a Byzantine is $\sum_{j=1}^{k} \alpha_j = 0.4$. Also assume that $P_{0,1}^k = P_{0,1}$ and $P_{1,0}^k = P_{1,0}, \forall k$. It can be seen that the optimal attack strategy comprises symmetric flipping probabilities and is given by $(P_{0,1}^k, P_{1,0}^k) = (1, 1)$, which corroborates the theoretical result presented in Lemma 3.13 and Theorem 3.5.

3.4.3 *Analysis from Network Designer's Perspective*

So far, the problem from the attacker's perspective has been discussed. Now let us look into the problem from a network designer's perspective and discuss techniques to mitigate the effect of the Byzantines.

Byzantine Identification Scheme

The authors in [15] extended the mitigation scheme proposed in [13] to identify the Byzantines in tree networks by assigning reputation to each node based on the quality of the data they provide. Assume that the FC has the knowledge of the attack model and utilizes this knowledge to identify the Byzantines. The FC observes the local decisions of each node over a time window T, which can be denoted by $(k, i) = [u_1(k, i), \ldots, u_T(k, i)]$ for $1 \leq i \leq N_k$ at level $1 \leq k \leq K$. Also assume that there is one honest anchor node with probability of detection P_d^A and probability of false alarm P_{fa}^A present and known to the FC. Let us denote the Hamming distance between the reports of the anchor node and an honest node i at level k over the time window T by $d_H^A(k, i) = ||U^A - U^H(k, i)||$, that is the number of elements that are different between U^A and $U^H(k, i)$. Similarly, the Hamming distance between the reports of the anchor node and a Byzantine node i at level k over the time window T is denoted by $d_B^A(k, i) = ||U^A - U^B(k, i)||$. Since the FC is aware of the fact that Byzantines might be present in the network, it compares the Hamming distance of a node i at level k to a threshold η_k, $\forall i, \forall k$, to make a decision to identify the Byzantines. In tree networks, a Byzantine node alters its decision as well as received decisions from its children prior to transmission in order to undermine the network performance. Therefore, solely based on the observed data of a node i at level k, the FC cannot determine whether the data has been flipped by the node i itself or by one of its Byzantine parent node. In this scheme, the FC makes the inference about a node being Byzantine by analyzing the data from the node i as well as the data from its predecessor nodes. FC starts from the nodes at level 1 and computes the Hamming distance between reports of the anchor node and the nodes at level 1. FC declares node i at level 1 to be a Byzantine if and only if the Hamming distance of node i is greater than a fixed threshold η_1. Children of identified Byzantine nodes $\mathbb{C}(\mathbb{B}_1)$ are not tested further because of the non-overlapping condition. However, if a level 1 node is determined not to be a Byzantine, then the FC tests its children nodes at level 2. The FC declares node i at level k, for $2 \leq k \leq K$, to be a Byzantine if and only if the Hamming distance of node i is greater than a fixed threshold η_k and Hamming distances of all predecessors of node i is less than equal to their respective thresholds η_j.

In this way, it is possible to counter the data falsification attack by isolating Byzantine nodes from the information fusion process. The probability that a Byzantine node i at level k is isolated at the end of the time window T is denoted as $P_B^{iso}(k, i)$.

Numerical Results

To gain insights into the solution, Fig. 3.18 presents some numerical results. Consider a tree network with $K = 5$ and plot $P_B^{iso}(k, i)$, $1 \leq k \leq 5$, as a function of the time window T. Assume that the operating points (P_d^k, P_{fa}^k), $1 \leq k \leq 5$, for the nodes at different levels are given by $[(0.8, 0.1), (0.75, 0.1), (0.6, 0.1), (0.65, 0.1), (0.6, 0.1)]$ and for anchor node $(P_d^A, P_{fa}^A) = (0.9, 0.1)$. We also assume that the hypotheses are equi-probable, i.e., $P_0 = P_1 = 0.5$, and the maximum isolation probability of honest nodes at level k based solely on its data is constrained

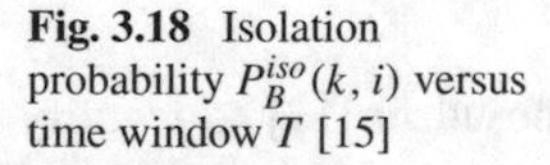

Fig. 3.18 Isolation probability $P_B^{iso}(k, i)$ versus time window T [15]

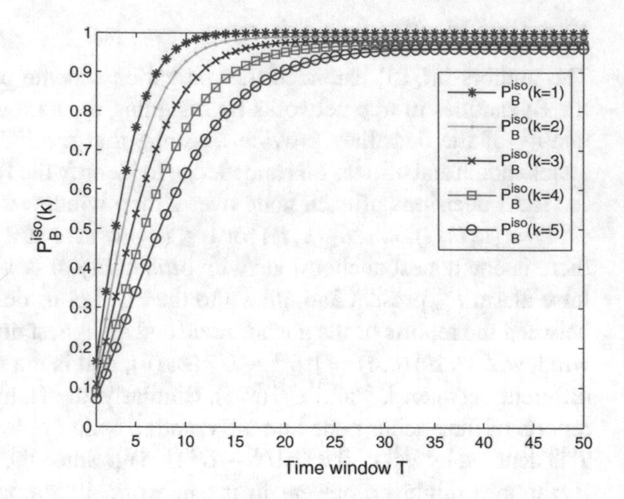

by $\delta_k = 0.01, \forall k$. It can be seen from Fig. 3.18 that in a span of only $T = 25$ time windows, the Byzantine identification scheme isolates/identifies almost all the Byzantines in the tree network.

3.5 Distributed Detection with Byzantine Data: Peer-to-Peer Networks

The traditional distributed detection framework discussed earlier comprises a group of spatially distributed nodes which acquire observations regarding the phenomenon of interest and send a compressed version to the fusion center (FC) where a global decision is made. However, in many scenarios, a centralized FC may not be available or the FC may become an information bottleneck causing degradation of system performance, potentially leading to system failure. Also, due to the distributed nature of future communication networks, and various practical constraints, e.g., absence of an FC, transmit power or hardware constraints and dynamic nature of the wireless medium, it may be desirable to employ alternate peer-to-peer local information exchange in order to reach a global decision. One such distributed approach for peer-to-peer local information exchange and inference is the use of a consensus algorithm [23]. The authors in [18] considered the problem of signal detection in peer-to-peer networks in the presence of data falsification (Byzantine) attacks. Detection approaches considered were based on fully decentralized consensus algorithms, where all of the nodes exchange information only with their neighbors in the absence of an FC. This architecture is often referred to as a decentralized architecture.

3.5.1 *System Model*

Network Model

Consider a network topology modeled as an undirected graph $G = (V, E)$, where $V = \{v_1, \cdots, v_N\}$ represents the set of nodes in the network with $|V| = N$. The set of communication links in the network correspond to the set of edges E, where $(v_i, v_j) \in E$, if and only if there is a communication link between v_i and v_j so that, v_i and v_j can directly communicate with each other. The adjacency matrix A of the graph is defined as

$$a_{ij} = \begin{cases} 1 \text{ if } (v_i, v_j) \in E, \\ 0 \text{ otherwise.} \end{cases} \tag{3.65}$$

The neighborhood of a node i is defined as

$$\mathcal{N}_i = \{v_j \in V : (v_i, v_j) \in E\}, \forall i \in \{1, 2, \cdots, N\}.$$

The degree d_i of a node v_i is the number of edges in E which include v_i as an endpoint, i.e., $d_i = \sum_{j=1}^{N} a_{ij}$.

The degree matrix D is defined as a diagonal matrix with $\text{diag}(d_1, \cdots, d_N)$ and the Laplacian matrix L is defined as

$$l_{ij} = \begin{cases} d_i \text{ if } j = i, \\ -a_{ij} \text{ otherwise.} \end{cases} \tag{3.66}$$

In other words, $L = D - A$. As an illustration, consider a network with six nodes trying to reach consensus (see Fig. 3.19). The Laplacian matrix L for this network is given by

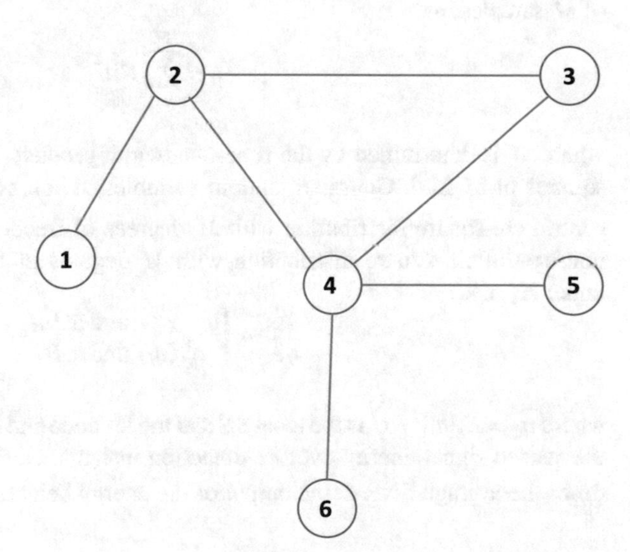

Fig. 3.19 A distributed network with six nodes

$$L = \begin{bmatrix} 1 & -1 & 0 & 0 & 0 & 0 \\ -1 & 3 & -1 & -1 & 0 & 0 \\ 0 & -1 & 2 & -1 & 0 & 0 \\ 0 & -1 & -1 & 4 & -1 & -1 \\ 0 & 0 & 0 & -1 & 1 & 0 \\ 0 & 0 & 0 & -1 & 0 & 1 \end{bmatrix}. \tag{3.67}$$

The consensus-based distributed detection scheme usually contains three phases: sensing, information fusion, and decision making. In the sensing phase, each node acquires the summary statistic about the phenomenon of interest. We adopt the energy detection method so that the local summary statistic is the received signal energy. Next, in the information fusion phase, each node communicates with its neighbors to update their state values (summary statistic) and continues with the consensus iteration until the whole network converges to a steady state which is the global test statistic. Finally, in the decision-making phase, nodes make their own decisions about the presence of the phenomenon using this global test statistic. In the following, each of these phases is described in more detail.

Sensing Phase

Consider an N-node network using the energy detection scheme [6]. For the ith node, the sensed signal z_i^t at time instant t is given by

$$z_i^t = \begin{cases} n_i^t, & \text{under } H_0 \\ \zeta_i s^t + n_i^t & \text{under } H_1, \end{cases} \tag{3.68}$$

where ζ_i is the deterministic gain corresponding to the sensing channel, s^t is the deterministic signal at time instant t, n_i^t is AWGN, i.e., $n_i^t \sim N(0, \sigma_i^2)$ and independent across time. Each node i calculates a summary statistic Y_i over a detection interval of M samples, as

$$Y_i = \sum_{t=1}^{M} |z_i^t|^2 \tag{3.69}$$

where M is determined by the time-bandwidth product. Since Y_i is the sum of the squares of M i.i.d. Gaussian random variables, it can be shown that $\frac{Y_i}{\sigma_i^2}$ follows a central chi-square distribution with M degrees of freedom (χ_M^2) under H_0, and, a non-central chi-square distribution with M degrees of freedom and parameter η_i under H_1, i.e.,

$$\frac{Y_i}{\sigma_i^2} \sim \begin{cases} \chi_M^2, & \text{under } H_0 \\ \chi_M^2(\eta_i) & \text{under } H_1 \end{cases} \tag{3.70}$$

where $\eta_i = E_s|h_i|^2/\sigma_i^2$ is the local SNR at the ith node and $E_s = \sum_{t=1}^{M} |s^t|^2$ represents the sensed signal energy over M detection instants. Note that the local SNR is M times the average SNR at the output of the energy detector, which is $\frac{E_s|h_i|^2}{M\sigma_i^2}$.

Information Fusion Phase

First, let us discuss conventional consensus algorithms [23] and see how consensus is reached using the following two steps.

Step 1: All nodes establish communication links with their neighbors, and broadcast their information state, $x_i(0) = Y_i$.

Step 2: Each node updates its local state information by a local fusion rule (weighted combination of its own value and those received from its neighbors). We denote node i's updated information at iteration k by $x_i(k)$. Node i continues to broadcast information $x_i(k)$ and updates its local information state until consensus is reached. This process of updating information state can be written in a compact form as

$$x_i(k+1) = x_i(k) + \frac{\varepsilon}{w_i} \sum_{j \in \mathcal{N}_i} (x_j(k) - x_i(k)) \tag{3.71}$$

where ε is the time step and w_i is the weight given to node i's information. Using the notation $x(k) = [x_1(k), \cdots, x_N(k)]^T$, network dynamics can be represented in the matrix form as,

$$x(k+1) = Wx(k) \tag{3.72}$$

where $W = I - \varepsilon \operatorname{diag}(1/w_1, \cdots, 1/w_N)L$ is referred to as a Perron matrix. The consensus algorithm is nothing but a local fusion or update rule that fuses the nodes' local information state with information coming from neighbor nodes, and it is well known that every node asymptotically reaches the same information state for arbitrary initial values [23].

Decision-Making Phase

The final information state x_i^* after reaching consensus for the above consensus algorithm will be the weighted average of the initial states of all the nodes [23] or $x_i^* = \sum_{i=1}^N w_i Y_i / \sum_{i=1}^N w_i$, $\forall i$. Average consensus can be seen as a special case of weighted average consensus with $w_i = w, \ \forall i$. After the whole network reaches a consensus, each node makes its own decision about the hypothesis using a predefined threshold λ[14]

$$\text{Decision} = \begin{cases} H_1 \text{ if } x_i^* > \lambda \\ H_0 \text{ otherwise} \end{cases} \tag{3.73}$$

where weights are given by [38]

$$w_i = \frac{\eta_i/\sigma_i^2}{\sum_{i=1}^N \eta_i/\sigma_i^2}. \tag{3.74}$$

$\Lambda = \sum_{i=1}^N w_i Y_i / \sum_{i=1}^N w_i$ is referred to as the final test statistic.

[14] In practice, parameters such as threshold λ and consensus time step ε are set off-line based on well-known techniques [23].

3.5.2 Analysis from Attacker's Perspective

The authors in [18] considered Byzantine attacks on consensus-based detection schemes. They analyzed the performance degradation of the weighted average consensus-based detection algorithms due to these attacks and proposed countermeasures.

Attack Model

When there are no adversaries in the network, consensus can be reached to the weighted average of arbitrary initial values by having the nodes use the update strategy $x(k+1) = Wx(k)$ with an appropriate weight matrix W. However, suppose that instead of broadcasting the true summary statistic Y_i and applying the update strategy (3.71), some Byzantine nodes deviate from the prescribed strategies. Accordingly, Byzantines can attack in two ways: data falsification (nodes falsify their initial data or weight values) and consensus disruption (nodes do not follow the update rule given by (3.71)). More specifically, Byzantine node i can do the following

$$\textit{Data falsification:} \quad x_i(0) = Y_i + \Delta_i, \quad \text{or} \quad w_i \text{ is changed to } \tilde{w}_i$$

$$\textit{Consensus disruption:} \quad x_i(k+1) = x_i(k) + \frac{\varepsilon}{w_i} \sum_{j \in \mathcal{N}_i} (x_j(k) - x_i(k)) + u_i(k),$$

where $(\Delta_i, \tilde{w}_i)$ and $u_i(k)$ are introduced at the initialization step and at the update step k, respectively. The attack model considered above is extremely general and allows Byzantine node i to update its value in a completely arbitrary manner (via appropriate choices of $(\Delta_i, \tilde{w}_i)$, and $u_i(k)$, at each time step). An adversary carrying out consensus disruption attack has the objective to disrupt the consensus operation. However, consensus disruption attacks can be easily detected because of the nature of the attack. The identification of consensus disruption attackers has been investigated in the literature (e.g., see [24, 30]) where control theoretical techniques were developed to identify disruption attackers in a 'single' consensus iteration. However, these techniques cannot identify the data falsification attacker due to the philosophically different nature of the problem. Also, note that by knowing the existence of such an identification mechanism, a smart adversary will aim to disguise itself while degrading the detection performance. In contrast to disruption attackers, data falsification attackers are more capable and can manage to disguise themselves while degrading the detection performance of the network by falsifying their data. Susceptibility and protection of consensus strategies to data falsification attacks has received scant attention, and this was the focus of the work in [18]. Their main focus[15] was on the scenarios where an attacker carries out a data falsification attack by introducing $(\Delta_i, \tilde{w}_i)$ during initialization. In such scenarios, one can exploit the statistical

[15]The authors in [18] also come up with a robust distributed weighted average consensus algorithm which allows detection of consensus disruption attacks while mitigating the effect of data falsification attacks.

distribution of the initial values and devise techniques to mitigate the influence of Byzantines on the distributed detection system.

In data falsification attacks, attackers try to manipulate the final test statistic (i.e., $\Lambda = (\sum_{i=1}^{N} w_i Y_i)/(\sum_{i=1}^{N} w_i)$) in a manner so as to degrade the detection performance. We consider a network with N nodes that uses Eq. (3.71) for reaching consensus. Weight w_i, given to node i's data Y_i in the final test statistic, is controlled or updated by node i itself while carrying out the iteration in Eq. (3.71). So by falsifying initial values Y_i or weights w_i, the attackers can manipulate the final test statistic. Detection performance will be degraded because Byzantine nodes can always set a higher weight to their manipulated information. Thus, the final statistic's value across the whole network will be dominated by the local statistic of the Byzantine node that will lead to degraded detection performance.

Next, let us define a mathematical model for data falsification attackers and analyze the degradation in detection performance of the network when Byzantines falsify their initial values Y_i for fixed arbitrary weights $\tilde{w}_i$. The objective of Byzantines is to degrade the detection performance of the network by falsifying their data (Y_i, w_i). Let us assume that Byzantines is endowed with extra knowledge and know the true hypothesis. Under this assumption, we analyze the detection performance of the data fusion scheme which yields the maximum performance degradation that the Byzantines can cause, i. e., worst case detection performance. Let us consider the case when weights of the Byzantines have been tampered with by setting their value at $\tilde{w}_i$ and analyze the effect of falsifying the initial values Y_i. Now a mathematical model for a Byzantine attack is presented. Byzantines tamper with their initial values Y_i and send $\tilde{Y}_i$ such that the detection performance is degraded.

Under H_0:

$$\tilde{Y}_i = \begin{cases} Y_i + \Delta_i \text{ with probability } P_i \\ \quad\; Y_i \text{ with probability } (1 - P_i) \end{cases}$$

Under H_1:

$$\tilde{Y}_i = \begin{cases} Y_i - \Delta_i \text{ with probability } P_i \\ \quad\; Y_i \text{ with probability } (1 - P_i) \end{cases}$$

where P_i is the attack probability and Δ_i is a constant value which represents the attack strength, which is zero for honest nodes. As we show later, Byzantine nodes will use a large value of Δ_i so that the final statistic's value is dominated by the Byzantine node's local statistic leading to a degraded detection performance.

The authors in [18] used deflection coefficient [20] to characterize the security performance of the detection scheme due to its simplicity and its strong relationship with the global detection performance. Deflection coefficient of the global test statistic is defined as: $\mathscr{D}(\Lambda) = \dfrac{(\mu_1 - \mu_0)^2}{\sigma_{(0)}^2}$, where $\mu_k = \mathbb{E}[\Lambda | H_k],\ k = 0, 1$, is the conditional mean and $\sigma_{(k)}^2 = \mathbb{E}[(\Lambda - \mu_k)^2 | H_k],\ k = 0, 1$, is the conditional variance. The deflection coefficient is closely related to performance measures such as the Receiver Operating Characteristics (ROC) curve [20]. In general, the detection performance monotonically improves with an increasing value of the deflection coefficient. The critical power of the distributed detection network is defined as the minimum fraction of Byzantine nodes needed to make the deflection coefficient of the global test statistic equal to zero (in which case, we say that the network becomes *blind*) and denote it by α_{blind}. Assume that the communication between nodes is error-free and the network topology remains fixed during the whole consensus process and, therefore, consensus can be reached without disruption [23].

Blinding the Network

The authors in [18] derived a closed-form expression for the minimum fraction of Byzantine nodes needed to blind the consensus-based distributed detection network. Without loss of generality, let us assume that the nodes corresponding to the first N_1 indices $i = 1, \cdots, N_1$ are Byzantines and the remaining nodes corresponding to indices $i = N_1 + 1, \cdots, N$ are honest nodes. Let us define $w = [\tilde{w}_1, \cdots, \tilde{w}_{N_1}, w_{N_1+1} \cdots, w_N]^T$ and $\text{sum}(w) = \sum_{i=1}^{N_1} \tilde{w}_i + \sum_{i=N_1+1}^{N} w_i$. The condition for blinding is provided in Lemma 3.14.

Lemma 3.14 ([18]) *For data fusion schemes, the condition to blind the network or equivalently to make the deflection coefficient zero is given by*

$$\mu_0 = \sum_{i=1}^{N_1}\left[P_i\frac{\tilde{w}_i}{sum(w)}(M\sigma_i^2 + \Delta_i) + (1 - P_i)\frac{\tilde{w}_i}{sum(w)}(M\sigma_i^2)\right] + \sum_{i=N_1+1}^{N}\left[\frac{w_i}{sum(w)}(M\sigma_i^2)\right] \tag{3.75}$$

$$\mu_1 = \sum_{i=1}^{N_1}\left[P_i\frac{\tilde{w}_i}{sum(w)}((M + \eta_i)\sigma_i^2 - \Delta_i) + (1 - P_i)\frac{\tilde{w}_i}{sum(w)}((M + \eta_i)\sigma_i^2)\right] + \sum_{i=N_1+1}^{N}\left[\frac{w_i}{sum(w)}((M + \eta_i)\sigma_i^2)\right] \tag{3.76}$$

$$\sigma_{(0)}^2 = \sum_{i=1}^{N_1}\left(\frac{\tilde{w}_i}{sum(w)}\right)^2 [P_i(1 - P_i)\Delta_i^2 + 2M\sigma_i^4] + \sum_{i=N_1+1}^{N}\left(\frac{w_i}{sum(w)}\right)^2 [2M\sigma_i^4] \tag{3.77}$$

$$\sum_{i=1}^{N_1} \tilde{w}_i(2P_i\Delta_i - \eta_i\sigma_i^2) = \sum_{i=N_1+1}^{N} w_i\eta_i\sigma_i^2.$$

Proof The proof is given in Appendix A.12. □

Note that when $w_i = \tilde{w}_i = z, \eta_i = \eta, \sigma_i = \sigma, P_i = P, \Delta_i = \Delta,\ \forall i$, the blinding condition simplifies to $\frac{N_1}{N} = \frac{1}{2}\frac{\eta\sigma^2}{P\Delta}$. This condition indicates that by appropriately choosing attack parameters (P_i, Δ_i), an adversary needs less than 50% data falsifying Byzantines to make the deflection coefficient zero.

Transient Performance Analysis with Byzantines

The authors in [18] further analyzed the detection performance of the data fusion schemes, denoted as $x(t+1) = W^t x(0)$, as a function of consensus iteration t in the presence of Byzantines. For analytical tractability, assume that $P_i = P,\ \forall i$. Denote by w_{ji}^t the element of matrix W^t in the jth row and ith column. Using these notations, we calculate the probability of detection and the probability of false alarm at the jth node at consensus iteration t.

For sufficiently large M (in practice $M \geq 12$), the distribution of Byzantine's data $\tilde{Y}_i$ given H_k is a Gaussian mixture which comes from $\mathcal{N}((\mu_{1k})_i, (\sigma_{1k})_i^2)$ with probability $(1-P)$ and from $\mathcal{N}((\mu_{2k})_i, (\sigma_{2k})_i^2)$ with probability P, where $\mathcal{N}$ denotes the normal distribution and

$$(\mu_{10})_i = M\sigma_i^2,\ (\mu_{20})_i = M\sigma_i^2 + \Delta_i \tag{3.78}$$

$$(\mu_{11})_i = (M+\eta_i)\sigma_i^2,\ (\mu_{21})_i = (M+\eta_i)\sigma_i^2 - \Delta_i \tag{3.79}$$

$$(\sigma_{10})_i^2 = (\sigma_{20})_i^2 = 2M\sigma_i^4,\ \textit{and}\ (\sigma_{11})_i^2 = (\sigma_{21})_i^2 = 2(M+2\eta_i)\sigma_i^4. \tag{3.80}$$

Now, the probability density function (PDF) of $x_{ji}^t = w_{ji}^t \tilde{Y}_i$ conditioned on H_k can be derived as

$$f(x_{ji}^t|H_k) = (1-P)\phi(w_{ji}^t(\mu_{1k})_i, (w_{ji}^t(\sigma_{1k})_i)^2) \tag{3.81}$$

$$+ P\phi(w_{ji}^t(\mu_{2k})_i, (w_{ji}^t(\sigma_{2k})_i)^2) \tag{3.82}$$

where $\phi(x|\mu, \sigma^2)$ (for notational convenience denoted as $\phi(\mu, \sigma^2)$) is the PDF of $X \sim \mathcal{N}(\mu, \sigma^2)$ and $\phi(x|\mu, \sigma^2) = \frac{1}{\sigma\sqrt{2\pi}} e^{-(x-\mu)^2/2\sigma^2}$. Next, for clarity of exposition, let us first derive the results for a small network with two Byzantine nodes and one honest node. Due to the probabilistic nature of the Byzantine's behavior, it may behave as an honest node with a probability $(1-P)$. Let S denote the set of all combinations of such Byzantine strategies:

$$S = \{\{b_1, b_2\}, \{h_1, b_2\}, \{b_1, h_2\}, \{h_1, h_2\}\} \tag{3.83}$$

where by b_i we mean that Byzantine node i behaves as a Byzantine and by h_i we mean that Byzantine node i behaves as an honest node. Let $A_s \in U$ denote the indices of Byzantines behaving as an honest node in the strategy combination s, then, from (3.83) we have

$$U = \{A_1 = \{\}, A_2 = \{1\}, A_3 = \{2\}, A_4 = \{1, 2\}\} \tag{3.84}$$

$$U^c = \{A_1^c = \{1,2\}, A_2^c = \{2\}, A_3^c = \{1\}, A_4^c = \{\}\} \tag{3.85}$$

where {} is used to denote the null set. Let us use m_s to denote the cardinality of subset $A_s \in U$. Using these notations, one can generalize these results for any arbitrary N.

Lemma 3.15 ([18]) *The test statistic of node j at consensus iteration t, i.e.,* $\tilde{\Lambda}_j^t = \sum_{i=1}^{N_1} w_{ji}^t \tilde{Y}_i + \sum_{i=N_1+1}^{N} w_{ji}^t Y_i$ *is a Gaussian mixture with PDF*

$$f(\tilde{\Lambda}_j^t|H_k) = \sum_{A_s \in U} P^{N_1-m_s}(1-P)^{m_s}\phi\left((\mu_k)_{A_s} + \sum_{i=N_1+1}^{N} w_{ji}^t(\mu_{1k})_i, \sum_{i=1}^{N}(w_{ji}^t(\sigma_{1k})_i)^2)\right)$$

$$\textit{with } (\mu_k)_{A_s} = \sum_{u \in A_s} w_{ju}^t(\mu_{1k})_j + \sum_{u \in A_s^c} w_{ju}^t(\mu_{2k})_j.$$

The performance of the detection scheme in the presence of Byzantines can be represented in terms of the probability of detection and the probability of false alarm of the network.

Proposition 3.3 ([18]) *The probability of detection and the probability of false alarm of node j at consensus iteration t in the presence of Byzantines can be represented as*

$$P_d^t(j) = \sum_{A_s \in U} P^{N_1-m_s}(1-P)^{m_s} Q\left(\frac{\lambda - (\mu_1)_{A_s} - \sum_{i=N_1+1}^{N} w_{ji}^t(\mu_{11})_i}{\sqrt{\sum_{i=1}^{N}(w_{ji}^t(\sigma_{11})_i)^2)}}\right) \textit{ and}$$

$$P_f^t(j) = \sum_{A_s \in U} P^{N_1-m_s}(1-P)^{m_s} Q\left(\frac{\lambda - (\mu_0)_{A_s} - \sum_{i=N_1+1}^{N} w_{ji}^t(\mu_{10})_i}{\sqrt{\sum_{i=1}^{N}(w_{ji}^t(\sigma_{10})_i)^2)}}\right),$$

where λ *is the threshold used for detection by node j.*

Numerical Results

Consider the six-node network with the topology given by the undirected graph shown in Fig. 3.19 deployed to detect a phenomenon. Nodes 1 and 2 are considered to be Byzantines. Sensing channel gains of the nodes are assumed to be $h = [0.8, 0.7, 0.72, 0.61, 0.69, 0.9]$ and weights are given by (3.74). Also assume that $M = 12$, $E_s = 5$, and $\sigma_i^2 = 1$, $\forall i$. Figure 3.20 shows the deflection coefficient of the global test statistic as a function of attack parameters $P_i = P, \Delta_i = \Delta, \forall i$. Note that the deflection coefficient is zero when the condition in Lemma 3.14 is satisfied. Another observation to make is that the deflection coefficient can be made zero even when only two out of six nodes are Byzantines. Thus, by appropriately choosing attack parameters (P, Δ), less than 50% of data falsifying Byzantines can blind the network.

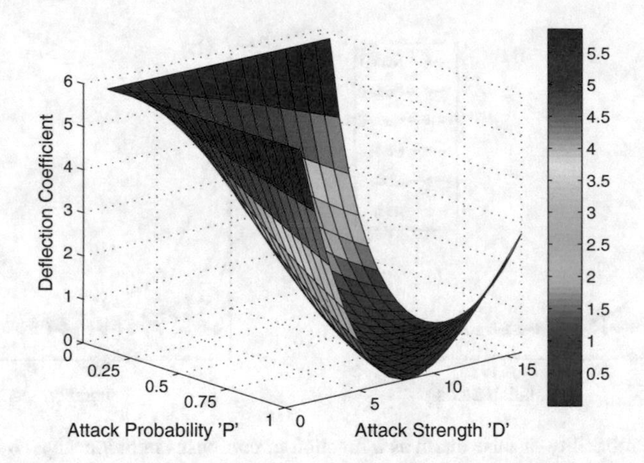

Fig. 3.20 Deflection coefficient as a function of attack parameters P and D [18]

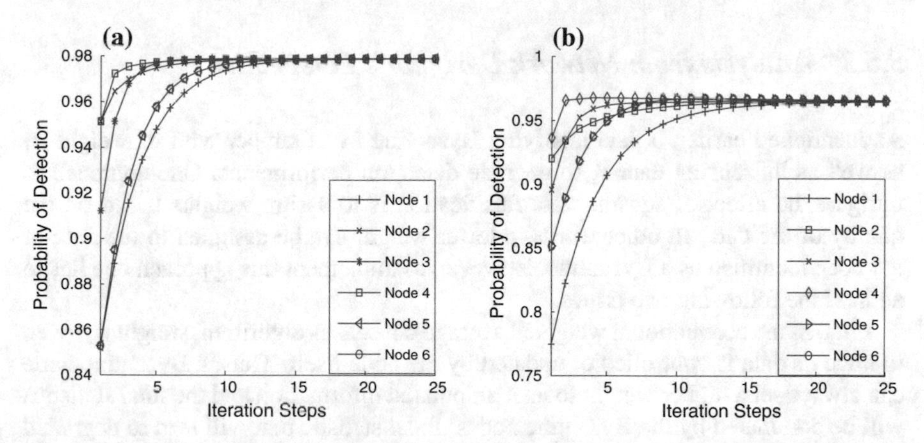

Fig. 3.21 **a** Probability of detection as a function of consensus iteration steps. **b** Probability of detection as a function of consensus iteration steps with Byzantines [18]

Next assume that $\varepsilon = 0.6897, \eta_i = 10, \sigma_i^2 = 2, \lambda = 33$ and $w_i = 1$. Attack parameters are assumed to be $(P_i, \Delta_i) = (0.5, 6)$ and $\tilde{w}_i = 1.1$. To characterize the transient performance of the weighted average consensus algorithm, Fig. 3.21a shows the probability of detection as a function of the number of consensus iterations without Byzantines, i.e., $(\Delta_i = 0, \tilde{w}_i = w_i)$. Next, Fig. 3.21b shows the probability of detection as a function of the number of consensus iterations in the presence of Byzantines. It can be seen that the detection performance degrades in the presence of Byzantines. Figure 3.22a shows the probability of false alarm as a function of the number of consensus iterations without Byzantines, i.e., $(\Delta_i = 0, \tilde{w}_i = w_i)$. Next, Fig. 3.22b shows the probability of false alarm as a function of the number of consensus iterations in the presence of Byzantines. From both Figs. 3.21 and 3.22, it can be seen that the Byzantine attack can severely degrade transient detection performance.

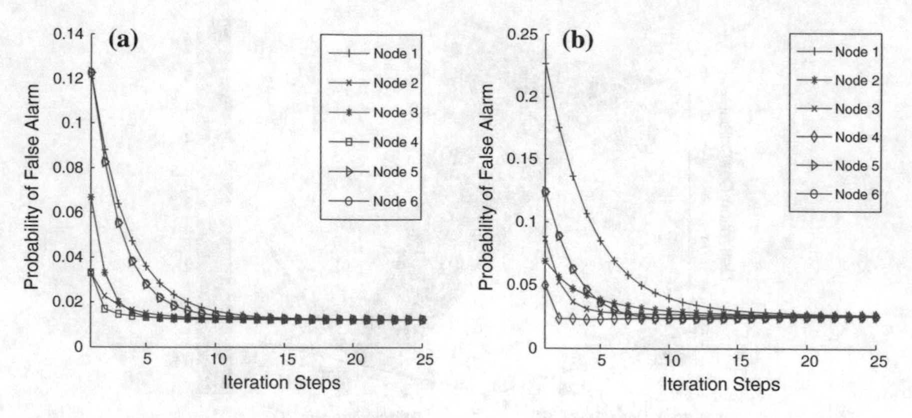

Fig. 3.22 **a** Probability of false alarm as a function of consensus iteration steps. **b** Probability of false alarm as a function of consensus iteration steps with Byzantines [18]

3.5.3 Analysis from Network Designer's Perspective

As mentioned earlier, a data falsifying Byzantine i can tamper with its weight w_i as well as its sensing data Y_i to degrade detection performance. One approach to mitigate the effect of sensing data falsification is to assign weights based on the quality of the data. In other words, a lower weight can be assigned to the data of the node identified as a Byzantine. However, to implement this approach one has to address the following two issues.

First, in the conventional weighted average consensus algorithm, weight w_i given to node i's data is controlled or updated by the node itself. Thus, a Byzantine node can always set a higher weight to its manipulated information and the final statistics will be dominated by the Byzantine nodes' local statistic that will lead to degraded detection performance. It will be impossible for any algorithm to detect this type of malicious behavior, since any weight that a Byzantine chooses for itself is a legitimate value that could also have been chosen by a node that is functioning correctly. Thus, in the conventional consensus algorithms weight manipulation cannot be detected, and therefore, conventional consensus algorithms cannot be used in the presence of an attacker.

Second, as will be seen later, the optimal weights given to nodes' sensing data depend on the following unknown parameters: identity of the nodes, which indicates whether the node is honest or Byzantine, and underlying statistical distribution of the nodes' data.

The authors in [18] proposed a learning-based robust weighted average consensus algorithm to address these concerns. More specifically, in order to address the first issue, which is the optimal weight design, they proposed a consensus algorithm in which the weight for node i's information is controlled (or updated) by the neighbors of node i rather than by node i itself. Note that networks deploying such an algorithm are more robust to weight manipulation attacks [18].

3.5.3.1 Distributed Algorithm for Weighted Average Consensus

The authors in [18] studied the following questions: does there exist a distributed algorithm that solves the weighted average consensus problem while satisfying the condition that weights must be controlled (or updated) by neighbors $\mathcal{N}_i$ of node i rather than by node i itself? If such an algorithm exists, then what are the conditions or constraints for the algorithm to converge?

Consider a network of N nodes with a fixed and connected topology $G(V, E)$. Next, let us state Perron–Frobenius theorem [10], which will be used later for the design and analysis of the robust weighted average consensus algorithm.

Theorem 3.6 ([18]) *Let W be a primitive nonnegative matrix with left and right eigenvectors u and v, respectively, satisfying $Wv = v$ and $u^T W = u^T$. Then $\lim_{k\to\infty} W^k = \frac{vu^T}{v^T u}$.*

Using Theorem 3.6, the authors in [18] adopted a reverse-engineering approach to design a modified Perron matrix $\hat{W}$ which has the weight vector $w = [w_1, w_2, \cdots, w_N]^T, w_i > 0, \forall i$ as its left eigenvector and $\mathbf{1}$ as its right eigenvector corresponding to eigenvalue 1. From the above theorem, if the modified Perron matrix $\hat{W}$ is primitive and nonnegative, then a weighted average consensus can be achieved. Now, the problem boils down to designing such a $\hat{W}$ which meets the requirement that weights are controlled (or updated) by the neighbors $\mathcal{N}_i$ of node i rather than by node i itself.

For this purpose, the authors in [18] proposed a modified Perron matrix $\hat{W} = I - \varepsilon(T \otimes L)$ where L is the original graph Laplacian, $\otimes$ is element-wise matrix multiplication operator, and T is a transformation given by

$$[T]_{ij} = \begin{cases} \frac{\sum\limits_{j\in\mathcal{N}_i} w_j}{l_{ii}} & \text{if } i = j \\ w_j & \text{otherwise.} \end{cases} \tag{3.86}$$

Observe that the above transformation T satisfies the condition that weights are controlled (or updated) by neighbors $\mathcal{N}_i$ of node i rather than by node i itself. Based on the above transformation T, the distributed consensus algorithm can be derived as:

$$x_i(k+1) = x_i(k) + \varepsilon \sum_{j\in\mathcal{N}_i} w_j (x_j(k) - x_i(k)).$$

Note that the form of this update equation is different from the conventional update equation. Let us denote the modified Perron matrix by $\hat{W} = I - \varepsilon\hat{L}$, where $\hat{L} = T \otimes L$. The convergence properties of the above algorithm were also established in [18]. It was shown that the consensus is reached asymptotically under certain conditions with value being $x_i^* = \frac{\sum_{i=1}^N w_i x_i(0)}{\sum_{i=1}^n w_i}, \forall i$.

In the discussed consensus algorithm, weights given to node i's data are updated by neighbors of the node i rather than by node i itself which addresses the first issue.

Adaptive Design of the Update Rules based on Learning of Nodes' Behavior

Next, to address the second issue, one can exploit the statistical distribution of the sensing data and devise techniques to mitigate the influence of Byzantines on the distributed detection system. The authors in [18] proposed a three-tier mitigation scheme where the following three steps are performed at each node: (1) identification of Byzantine neighbors, (2) estimation of parameters of identified Byzantine neighbors, and (3) adaptation of consensus algorithm (or update weights) using estimated parameters.

Let us first discuss the design of distributed optimal weights for the honest/Byzantine nodes assuming that the identities of the nodes are known. Later, we will explain how the identity of nodes (i.e., honest/Byzantine) can be determined.

(1) Design of Distributed Optimal Weights in the Presence of Byzantines

Let us denote by δ_i^B, the centralized weight given to the Byzantine node and by δ_i^H, the centralized weight given to the Honest node. By considering $\delta_i^B = w_i^B / \left(\sum_{i=1}^{N_1} w_i^B + \sum_{i=N_1+1}^{N} w_i^H \right)$ and $\delta_i^H = w_i^H / \left(\sum_{i=1}^{N_1} w_i^B + \sum_{i=N_1+1}^{N} w_i^H \right)$, the optimal weight design problem which maximizes the deflection coefficient can be stated formally as:

$$\begin{aligned} &\max_{\{\delta_i^B\}_{i=1}^{N_1}, \{\delta_i^H\}_{i=N_1+1}^{N}} \frac{(\mu_1 - \mu_0)^2}{\sigma_{(0)}^2} \\ &\text{s.t.} \quad \sum_{i=1}^{N_1} \delta_i^B + \sum_{i=N_1+1}^{N} \delta_i^H = 1 \end{aligned} \tag{P3.7}$$

where μ_1, μ_0, and $\sigma_{(0)}^2$ are given in (3.75), (3.76) and (3.77), respectively. The solution of the Problem P3.7 is presented in the next lemma.

Lemma 3.16 ([18]) *Optimal centralized weights which maximize the deflection coefficient are given as*

$$\delta_i^B = \frac{w_i^B}{\sum_{i=1}^{N_1} w_i^B + \sum_{i=N_1+1}^{N} w_i^H},$$

$$\delta_i^H = \frac{w_i^H}{\sum_{i=1}^{N_1} w_i^B + \sum_{i=N_1+1}^{N} w_i^H}$$

where $w_i^B = \dfrac{(\eta_i \sigma_i^2 - 2P_i \Delta_i)}{\Delta_i^2 P_i (1 - P_i) + 2M\sigma_i^4}$ *and* $w_i^H = \dfrac{\eta_i}{2M\sigma_i^2}$.

Remark 3.1 Distributed optimal weights can be chosen as w_i^B and w_i^H. Thus, the value of the global test statistic (or final weighted average consensus) is the same as the optimal centralized weighted combining scheme.

Note that the optimal weights for the Byzantines are functions of the attack parameters $(P_i,\ \Delta_i)$, which may not be known to the neighboring nodes in practice. In addition, the parameters of the honest nodes might also not be known. Therefore, [18] proposed a technique to learn or estimate these parameters and, then, use these estimates to adaptively design the local fusion rule which are updated after each learning iteration.

(2) Identification, Estimation, and Adaptive Fusion Rule

The first step at each node m is to determine the identity ($I^i \in \{H, B\}$) of its neighboring nodes $i \in \mathcal{N}_m$. Note that if node i is an honest node, its data under hypothesis H_k is normally distributed $\mathcal{N}((\mu_{1k})_i,\ (\sigma_{1k})_i^2)$. On the other hand, if node i is a Byzantine node, its data under hypothesis H_k is a Gaussian mixture which comes from $\mathcal{N}((\mu_{1k})_i,\ (\sigma_{1k})_i^2)$ with probability ($\alpha_1^i = 1 - P_i$) and from $\mathcal{N}((\mu_{2k})_i,\ (\sigma_{2k})_i^2)$ with probability $\alpha_2^i = P_i$. Therefore, determining the identity ($I^i \in \{H, B\}$) of neighboring nodes $i \in \mathcal{N}_m$ can be posed as a hypothesis testing problem:

$\mathbf{I_0}\ (\mathbf{I^i = H}):$ Y_i is generated from a Gaussian distribution under each hypothesis H_k;

$\mathbf{I_1}\ (\mathbf{I^i = B}):$ Y_i is generated from a Gaussian mixture distribution under each hypothesis H_k.

Node classification can then be achieved using the maximum likelihood decision rule:

$$f(Y_i|\ I_0) \ \underset{B}{\overset{H}{\gtrless}}\ f(Y_i|\ I_1) \tag{3.87}$$

where $f(Y_i \mid I_l)$ is the probability density function (PDF) of Y_i under each hypothesis I_l. However, the parameters of the distributions are not known. For an honest node i, the parameters to be estimated are $((\mu_{1k})_i,\ (\sigma_{1k})_i^2)$, and for Byzantines, the unknown parameter set to be estimated is $\theta = \{\alpha_j^i,\ (\mu_{jk})_i,\ (\sigma_{jk})_i^2\}$, where $k = \{0, 1\}, j = \{1, 2\}$ and $i = 1, \cdots, N_m$, for N_m neighboring nodes. The authors in [18] employed a technique based on the expectation–maximization (EM) algorithm and maximum likelihood (ML) estimation to learn these parameters. These parameters were estimated and updated by observing the data over multiple learning iterations. In each learning iteration t, each node in the network employs the data coming from their neighbors for D detection intervals to learn their respective parameters. It was assumed that each node has the knowledge of the true hypothesis for D detection intervals (or history) through a feedback mechanism.

Numerical Results

The effectiveness of the detection scheme is presented through numerical results. For this scenario, consider the six-node network shown in Fig. 3.19 where the nodes employ the discussed algorithm (with $\varepsilon = 0.3$) to reach a consensus. To gain insights into the convergence property of the consensus algorithm, Fig. 3.23 shows the updated state values at each node as a function of consensus iterations. Assume that the initial data vector is $x(0) = [5, 2, 7, 9, 8, 1]^T$ and the weight vector

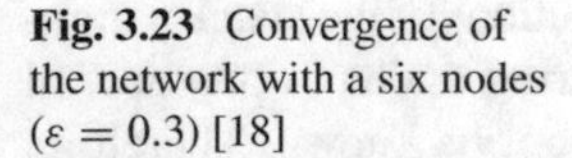

Fig. 3.23 Convergence of the network with a six nodes ($\varepsilon = 0.3$) [18]

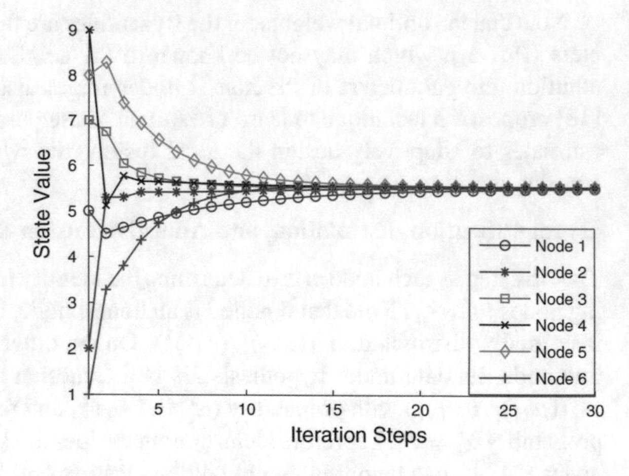

Fig. 3.24 ROC for different protection approaches [18]

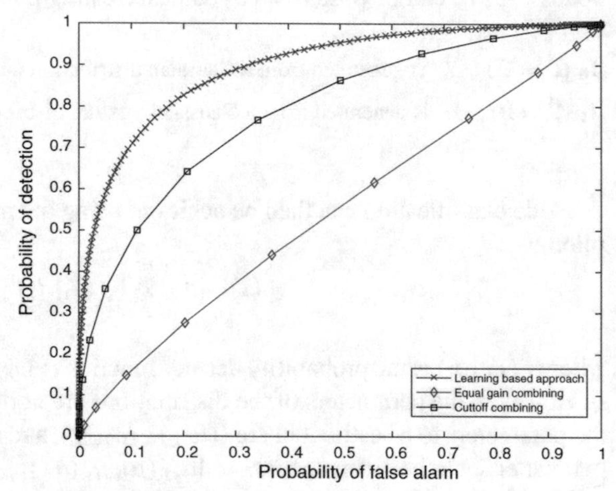

is $w = [0.65, 0.55, 0.48, 0.95, 0.93, 0.90]^T$. Figure 3.23 shows the convergence of the consensus algorithm iterations. It is observed that within 20 iterations, consensus is reached on the global decision statistics, the weighted average of the initial values (states).

Next, Figs. 3.24 and 3.25 present some numerical results to evaluate the performance of the discussed scheme with Byzantines. Node 1 and node 2 are considered to be Byzantines. Assume that $M = 12$, $\eta_i = 3$, $\sigma_i^2 = 0.5$ and the attack parameters are $(P_i, \Delta_i) = (0.5, 9)$. Figure 3.24 compares the discussed weighted average consensus-based detection scheme with the equal gain combining scheme[16] and the scheme where Byzantines are excluded from the fusion process. It can be clearly

[16]In the equal gain combining scheme, all the nodes (including Byzantines) are given the same weight.

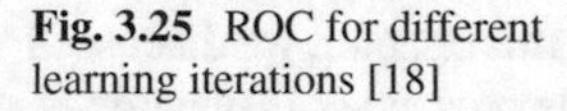
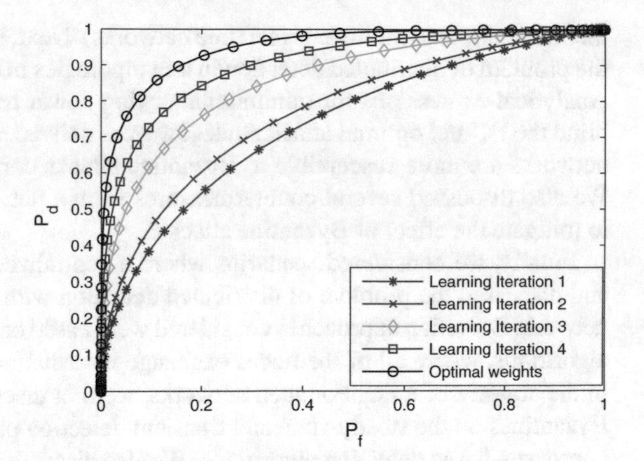

Fig. 3.25 ROC for different learning iterations [18]

seen from the figure that the learning-based scheme performs better than the rest of the schemes.

Next, assume that $((\mu_{10})_i, (\sigma_{10})_i^2) = (3, 1.5)$, $((\mu_{11})_i, (\sigma_{11})_i^2) = (4, 2)$ and the attack parameters are $(P_i, \Delta_i) = (0.5, 9)$. Figure 3.25 shows ROC curves for different number of learning iterations. For every learning iteration, we assume that $D_1 = 10$ and $D = 20$. It can be seen from Fig. 3.25 that within 4 learning iterations, detection performance of the learning-based weighted gain combining scheme approaches the detection performance of weighted gain combining with known optimal weight-based scheme.

Some recent research has also focused on using decentralized optimization-based solutions to design robust algorithms that counter Byzantines in peer-to-peer networks [19, 36].

3.6 Summary

In this chapter, we discussed distributed detection in the presence of Byzantines. First, we discussed the case of a parallel network performing a detection task using binary quantized data. The expression for minimum attacking power required by the Byzantines to blind the FC was presented. We also discussed the problem under different practical scenarios where the FC and the Byzantines may or may not have knowledge of their opponent's strategies and presented results for both asymptotic and non-asymptotic cases. It was found that asymptotic-based results do not hold under several non-asymptotic scenarios. We also discussed the problem of distributed detection with parallel network topology in a Neyman–Pearson framework. Optimal attack strategies and some Byzantine mitigation schemes were discussed.

In several practical situations, however, a parallel topology cannot be implemented. In such cases, a multi-hop network is employed, where nodes are organized hierarchically into multiple levels (tree networks). Next, in this chapter, we discussed the problem of distributed detection in tree topologies in the presence of Byzantines. Analytical expressions for minimum attacking power required by the Byzantine to blind the FC and optimal attack strategies were derived. It was found that multi-hop networks are more susceptible to Byzantine attacks compared to parallel network. We also discussed several countermeasures from a network designer's perspective to mitigate the effect of Byzantine attacks.

Finally, we considered scenarios where a centralized FC may not be available and discussed the problem of distributed detection with Byzantines in peer-to-peer network. Detection approaches considered were based on fully distributed consensus algorithms, where all of the nodes exchange information only with their neighbors in the absence of a FC. For such networks, we first discussed the negative effect of Byzantines on the steady-state and transient detection performance of conventional consensus-based detection algorithms. We also discussed mitigation schemes which are robust to Byzantine attacks.

References

1. Abrardo A, Barni M, Kallas K, Tondi B (2016) A game-theoretic framework for optimum decision fusion in the presence of Byzantines. IEEE Trans Inf Forensics Secur 11(6):1333–1345. https://doi.org/10.1109/TIFS.2016.2526963
2. Bard JF (1991) Some properties of the bilevel programming problem. J Optim Theory Appl 68(2). https://doi.org/10.1007/BF00941574
3. Chair Z, Varshney P (1986) Optimal data fusion in multiple sensor detection systems. IEEE Trans Aerosp Electron Syst AES-22(1):98–101
4. Chen H, Varshney LR, Varshney PK (2014) Noise-enhanced information systems. Proc IEEE 102(10):466–471. https://doi.org/10.1109/TSP.2009.2028938
5. Deineko VG, Woeginger GJ (2011) A well-solvable special case of the bounded knapsack problem. Oper Res Lett 39:118–120
6. Digham F, Alouini MS, Simon MK (2003) On the energy detection of unknown signals over fading channels. In: IEEE international conference on communications, 2003. ICC'03, vol 5, pp 3575–3579. https://doi.org/10.1109/ICC.2003.1204119
7. Gagrani M, Sharma P, Iyengar S, Nadendla VSS, Vempaty A, Chen H, Varshney PK (2011) On noise-enhanced distributed inference in the presence of Byzantines. In: Proceedings of the 49th annual allerton conference on communication, control, and computing, pp 1222–1229. https://doi.org/10.1109/Allerton.2011.6120307
8. Hashlamoun W, Brahma S, Varshney PK (2018a) Audit bit based distributed Bayesian detection in the presence of Byzantines. https://doi.org/10.1109/TSIPN.2018.2806842
9. Hashlamoun W, Brahma S, Varshney PK (2018b) Mitigation of Byzantine attacks on distributed detection systems using audit bits. IEEE Trans Signal Inf Process over Nets 4(1):18–32. https://doi.org/10.1109/TSIPN.2017.2723723
10. Horn RA, Johnson CR (1987) Matrix Analysis. Cambridge University Press, Cambridge, UK
11. Kailkhura B, Brahma S, Han YS, Varshney PK (2013a) Optimal distributed detection in the presence of Byzantines. In: Proceedings of the 38th international conference on acoustics, speech, and signal processing (ICASSP 2013). Vancouver, Canada
12. Kailkhura B, Brahma SK, Han YS, Varshney PK (2013b) Optimal distributed detection in the presence of Byzantines. In: Proceedings of the IEEE international conference acoustics, speech, and signal processing (ICASSP 2013), pp 2925–2929. https://doi.org/10.1109/ICASSP.2013.6638193

13. Kailkhura B, Han YS, Brahma S, Varshney PK (2013c) On covert data falsification attacks on distributed detection systems. In: 2013 13th international symposium on communications and information technologies (ISCIT). IEEE, pp 412–417
14. Kailkhura B, Brahma S, Han YS, Varshney PK (2014) Distributed detection in tree topologies with Byzantines. IEEE Trans Signal Process 12(62):3208–3219
15. Kailkhura B, Brahma S, Dulek B, Han YS, Varshney PK (2015a) Distributed detection in tree networks: Byzantines and mitigation techniques. IEEE Trans Inf Forensics Secur 10(7):1499–1512
16. Kailkhura B, Han YS, Brahma S, Varshney PK (2015b) Asymptotic analysis of distributed bayesian detection with Byzantine data. IEEE Signal Process Lett 22(5):608–612
17. Kailkhura B, Han YS, Brahma S, Varshney PK (2015c) Distributed bayesian detection in the presence of Byzantine data. IEEE Trans Signal Process 63(19):5250–5263
18. Kailkhura B, Brahma S, Varshney PK (2017a) Data falsification attacks on consensus-based detection systems. IEEE Trans Signal Inf Process Netw 3(1):145–158
19. Kailkhura B, Ray P, Rajan D, Yen A, Barnes P, Goldhahn R (2017b) Byzantine-resilient locally optimum detection using collaborative autonomous networks. In: 2017 IEEE 7th international workshop on computational advances in multi-sensor adaptive processing (CAMSAP), pp 1–5. https://doi.org/10.1109/CAMSAP.2017.8313059
20. Kay SM (1998) Fundamentals of statistical signal processing: detection theory. Prentice hall signal processing series, vol. 2. A. V. Oppenheim, Ed. Prentice Hall PTR
21. Lin Y, Chen B, Varshney P (2005) Decision fusion rules in multi-hop wireless sensor networks. IEEE Trans Aerosp Electron Syst 41(2):475–488. https://doi.org/10.1109/TAES.2005.1468742
22. Marano S, Matta V, Tong L (2009) Distributed detection in the presence of Byzantine attacks. IEEE Trans Signal Process 57(1):16–29. https://doi.org/10.1109/TSP.2008.2007335
23. Olfati-Saber R, Fax J, Murray R (2007) Consensus and cooperation in networked multi-agent systems. Proc IEEE 95(1):215–233
24. Pasqualetti F, Bicchi A, Bullo F (2012) Consensus computation in unreliable networks: a system theoretic approach. IEEE Trans Autom Control 57(1):90–104
25. Rawat AS, Anand P, Chen H, Varshney PK (2011) Collaborative spectrum sensing in the presence of Byzantine attacks in cognitive radio networks. IEEE Trans Signal Process 59(2):774–786. https://doi.org/10.1109/TSP.2010.2091277
26. Shi W, Sun TW, Wesel RD (1999) Optimal binary distributed detection. In: Proceedings of the 33rd asilomar conference on signals, systems, and computers, pp 24–27
27. Soltanmohammadi E, Naraghi-Pour M (2014a) Fast detection of malicious behavior in cooperative spectrum sensing. IEEE J Sel Areas Commun 32(3):377–386. https://doi.org/10.1109/JSAC.2014.140301
28. Soltanmohammadi E, Naraghi-Pour M (2014b) Nonparametric density estimation, hypotheses testing, and sensor classification in centralized detection. IEEE Trans Inf Forensics Secur 9(3):426–435. https://doi.org/10.1109/TIFS.2014.2300939
29. Song WZ, Huang R, Shirazi B, LaHusen R (2009) TreeMAC: Localized TDMA MAC protocol for real-time high-data-rate sensor networks. In: IEEE international conference on pervasive computing and communications, 2009. PerCom 2009, pp 1–10. https://doi.org/10.1109/PERCOM.2009.4912757
30. Sundaram S, Hadjicostis C (2011) Distributed function calculation via linear iterative strategies in the presence of malicious agents. IEEE Trans Autom Control 56(7):1495–1508. https://doi.org/10.1109/TAC.2010.2088690
31. Tsitsiklis JN (1988) Decentralized detection by a large number of sensors. Signals Syst Math Control 1(2):167–182
32. Varshney PK (2012) Distributed detection and data fusion. Springer Science & Business Media
33. Vempaty A, Agrawal K, Varshney P, Chen H (2011) Adaptive learning of Byzantines' behavior in cooperative spectrum sensing. In: 2011 IEEE wireless communications and networking conference, pp 1310–1315. https://doi.org/10.1109/WCNC.2011.5779320

34. Vempaty A, Nadendla VSS, Varshney PK (2013) Further results on noise-enhanced distributed inference in the presence of Byzantines. In: Proceedings of the 16th international symposium wireless personal multmedia communications (WPMC), pp 1–5
35. Vempaty A, Ray P, Varshney PK (2014) False discovery rate based distributed detection in the presence of Byzantines. IEEE Trans Aerosp Electron Syst 50(3):1826–1840. https://doi.org/10.1109/TAES.2014.120645
36. Yen AY, Barnes PD, Kailkhura B, Ray P, Rajan D, Schmidt KL, Goldhahn RA (2018) Large-scale parallel simulations of distributed detection algorithms for collaborative autonomous sensor networks. In: Proceedings of the SPIE, pp 1–7. https://doi.org/10.1117/12.2306545
37. Zhang Q, Varshney P, Wesel R (2002) Optimal bi-level quantization of i.i.d. sensor observations for binary hypothesis testing. IEEE Trans Inf Theory 48(7):2105–2111. https://doi.org/10.1109/TIT.2002.1013153
38. Zhang W, Wang Z, Guo Y, Liu H, Chen Y, Mitola J (2011) Distributed cooperative spectrum sensing based on weighted average consensus. In: Global telecommunications conference (GLOBECOM 2011), 2011. IEEE, pp 1–6. https://doi.org/10.1109/GLOCOM.2011.6134149

Chapter 4
Distributed Estimation and Target Localization

In this chapter, we discuss another important inference problem, distributed parameter estimation, in the presence of Byzantines.

We first consider the specific problem of target localization. In the location estimation problem, sensors quantize their local observations and send it to the FC. The FC estimates the location of the target using the sensors' locations and their quantized observations. In such a system, a target localization scheme based on Monte Carlo methods is typically used for the Bayesian setup. We present the fundamental limits on attack strategies in this setup. When the problem is considered from the attacker's perspective, two attack strategies can be considered: independent and collaborative. For the inference task considered, the optimal attack is defined as the one that minimizes the posterior Fisher information or maximizes the posterior Cramér–Rao lower bound (PCRLB) as defined in Chap. 2. Several countermeasures including a recently introduced error-correcting codes-based mitigation strategy for localization are also presented.

We then consider a parameter estimation problem and present results in the asymptotic setup. We consider the case when several subsets of sensors are assumed to be attacked using different attacks and discuss a scheme to identify and categorize the attacked sensors into different groups as per the types of attacks. It is shown that as the number of time samples at each sensor and the number of sensors increase, the performance of the scheme will improve but to different extents. Next, in order to improve the estimation performance, the data of the identified sensors can be used. For this purpose, a joint estimation of the attack parameters and the unknown parameter is considered. When the quantization approach is not changed post-attack, it is shown that the corresponding Fisher information matrix (FIM) is singular. To overcome this, a time-varying quantization approach has been proposed, which will provide a non-singular FIM.

Finally, we consider distributed estimation with Byzantines in a peer-to-peer network. An algorithm termed Flag Raising Distributed Estimation (FRDE), is discussed that robustly estimates the parameter in the presence of Byzantines.

A. Vempaty et al., *Secure Networked Inference with Unreliable Data Sources*,
https://doi.org/10.1007/978-981-13-2312-6_4

4.1 Target Localization with Byzantine Data: Parallel Networks

4.1.1 System Model

Consider a scenario where N sensors are deployed in a WSN to estimate the location of a target present at $\theta = [x_t, y_t]$ where x_t and y_t denote the coordinates of the target location in the 2-D Cartesian plane as shown in Fig. 4.1. Although the sensors in Fig. 4.1 are shown to be deployed on a regular grid, the schemes discussed here are capable of handling any kind of sensor deployment as long as the location information for each sensor is available at the FC. Assume that the target location has a prior distribution $p_0(\theta)$. For simplicity, we assume that $p_0(\theta)$ is a Gaussian distribution, i.e., $\theta \sim \mathcal{N}(\mu, \sigma_\theta^2 \mathbf{I})$, where the mean μ is the center of a region of interest (ROI) and $\sigma_\theta^2 \mathbf{I}$ is very large such that the ROI includes the target's (100-t)% confidence region (typically taken to be 99%). The signal radiated from this location is assumed to follow an isotropic power attenuation model given as

$$a_i^2 = P_0 \left(\frac{d_0}{d_i} \right)^n, \tag{4.1}$$

where a_i is the signal amplitude received at the ith sensor, P_0 is the power measured at a reference distance d_0, n is the path-loss exponent, and d_i is the distance between the target and the ith sensor. Note that $d_i \neq 0$, i.e., the target is not colocated with

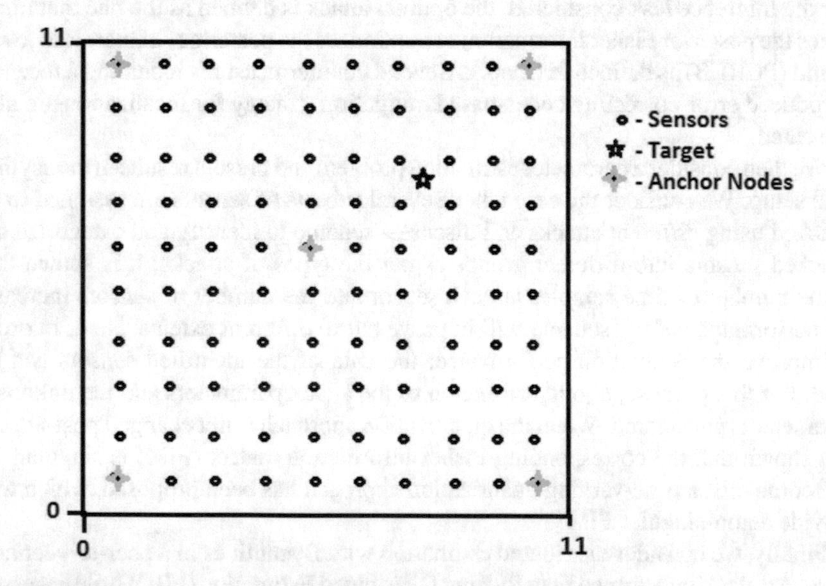

Fig. 4.1 Target in a grid deployed sensor field with anchor sensors

a sensor. This assumption is valid as the probability of a target being exactly at the same location as the sensor is zero; therefore, a_i is almost surely not unbounded. Without loss of generality, let $d_0 = 1$ and $n = 2$. The signal amplitude is assumed to be corrupted by additive white Gaussian noise (AWGN) at each sensor:

$$s_i = a_i + n_i, \tag{4.2}$$

where s_i is the corrupted signal at the ith sensor and the noise n_i follows $\mathcal{N}(0, \sigma^2)$. This noise is considered to be independent across sensors. Note that the signal model given in (4.1) and (4.2) has been verified experimentally for acoustic signals in [7] and it results from averaging time samples of the received acoustic energy. The interested reader is referred to [7, 9, 14] for details.

Due to bandwidth and energy limitations, each sensor uses a binary quantizer and sends its quantized binary measurement to the FC. The FC is assumed to know the sensor locations, and the sensors use threshold quantizers because of their simplicity [13] in terms of both implementation and analysis:

$$D_i = \begin{cases} 0 & s_i < \eta_i \\ 1 & s_i > \eta_i \end{cases} \tag{4.3}$$

where D_i is the quantized binary measurement and η_i is the quantization threshold at the ith sensor. The FC receives the binary vector $\mathbf{u} = [u_1, \ldots, u_N]$ from all the sensors in the network. After collecting $\mathbf{u}$, the FC estimates the location $\theta = [x_t, y_t]$ by using the minimum mean square error (MMSE) estimator as in [8], i.e., $\hat{\theta} = E[\theta|\mathbf{u}]$, where $E[\cdot|\mathbf{u}]$ denotes the expectation with respect to the posterior probability density function (PDF) $p(\theta|\mathbf{u})$. Since $E[\theta|\mathbf{u}]$ cannot be calculated in a closed form, an important sampling-based Monte Carlo method was used in [19]. It is also assumed that there are K anchor sensors in the network similar to [8]. These anchor sensors are assumed to be secure and are used to obtain an initial estimate of the target location, θ.

4.1.1.1 Monte Carlo Method-Based Target Localization

With the recent advances in computation power, Monte Carlo-based methods have become useful tools for inference problems. An important sampling-based Monte Carlo method [4] is used to approximate the posterior PDF, $p(\theta|\mathbf{u})$, as

$$p(\theta|\mathbf{u}) = \sum_{m=1}^{N_p} w_m \delta(\theta - \theta_m), \tag{4.4}$$

where the approximation is obtained as a weighted sum of N_p particles. The particles $\theta_m = [x_m, y_m]$ are drawn from the prior distribution $p_0(\theta)$. The weights are calculated using the data $\mathbf{u}$ and are proportional to the likelihood function

$$\tilde{w}_m \propto p(\mathbf{u}|\theta_m)w_m^0, \tag{4.5}$$

where the initial weights are set as identical, i.e., $w_m^0 = 1/N_p$. The updated weight of each particle is the original weight multiplied by the likelihood function of the data. Since the sensors' data is conditionally independent, we have $p(\mathbf{u}|\theta_m) = \prod_{i=1}^N p(u_i|\theta_m)$ for $m = 1, \ldots, N_p$. The particle weights are then normalized as

$$w_m = \frac{\tilde{w}_m}{\sum_{m=1}^{N_p} \tilde{w}_m} \tag{4.6}$$

$$= \frac{p(\mathbf{u}|\theta_m)}{\sum_{m=1}^{N_p} p(\mathbf{u}|\theta_m)} \tag{4.7}$$

This results in the location estimate $\hat{\theta}$ given by

$$\hat{\theta} = \sum_{m=1}^{N_p} w_m \theta_m. \tag{4.8}$$

4.1.1.2 Performance Metrics

As discussed in Sect. 2.1.4, PCRLB and posterior Fisher information matrix (FIM) are used as the performance metrics to analyze the estimation performance [15, 16]. Let $\hat{\theta}(\mathbf{u})$ be an estimator of the target location θ. From (2.15),

$$\mathbf{F} = -E_{\theta,\mathbf{u}}[\nabla_\theta \nabla_\theta^T \ln P(\mathbf{u}, \theta)] \tag{4.9}$$

$$= -E_{\theta,\mathbf{u}}[\nabla_\theta \nabla_\theta^T \ln P(\mathbf{u}|\theta)] - E_\theta[\nabla_\theta \nabla_\theta^T \ln p_0(\theta)] \tag{4.10}$$

$$= \mathbf{F_D} + \mathbf{F_P} \tag{4.11}$$

where ∇_θ is the gradient operator defined as $\nabla_\theta = \left[\frac{\partial}{\partial x_t}, \frac{\partial}{\partial y_t}\right]^T$. $\mathbf{F_D}$ and $\mathbf{F_P}$ represent the contributions of data and the prior to $\mathbf{F}$, respectively. The elements of $\mathbf{F}$ are:

$$F_{11} = \int p_0(\theta) \left(\sum_{i=1}^N \sum_{l=0}^1 \frac{1}{P(u_i = l|\theta)} \left[\frac{\partial P(u_i = l|\theta)}{\partial x_t} \right]^2 \right) d\theta + \frac{1}{\sigma_\theta^2}, \tag{4.12}$$

$$F_{22} = \int p_0(\theta) \left(\sum_{i=1}^N \sum_{l=0}^1 \frac{1}{P(u_i = l|\theta)} \left[\frac{\partial P(u_i = l|\theta)}{\partial y_t} \right]^2 \right) d\theta + \frac{1}{\sigma_\theta^2}, \tag{4.13}$$

$$F_{21} = F_{12} = \int p_0(\theta) \sum_{i=1}^{N} \sum_{l=0}^{1} \frac{1}{P(u_i = l|\theta)} \left[\frac{\partial P(u_i = l|\theta)}{\partial x_t} \right] \left[\frac{\partial P(u_i = l|\theta)}{\partial y_t} \right] d\theta, \tag{4.14}$$

where $P(u_i = l|\theta)$ for $l = 0, 1$ is the probability that a sensor sends $u_i = l$. This value depends on Byzantines' attacking strategy.

4.1.2 *Analysis from Attacker's Perspective*

Independent Attack Model

In [19], the independent Byzantine attack model was considered and the fundamental limits and mitigation strategies were designed. In an independent attack, each Byzantine sensor attacks the network by relying on its own observation without any knowledge regarding the presence of other Byzantines or their observations. Let the number of Byzantines present in the network be $M = \alpha N$. When the channels between the sensors and the FC are ideal, for an honest sensor, $u_i = D_i$, whereas for the Byzantines we assume that they flip their quantized binary measurements with probability p. Therefore, under the Gaussian noise assumption, the probability that $u_i = 1$ is given by

$$P(u_i = 1|\theta, i = Honest) = Q\left(\frac{\eta_i - a_i}{\sigma}\right) \tag{4.15}$$

$$P(u_i = 1|\theta, i = Byzantine) = p\left(1 - Q\left(\frac{\eta_i - a_i}{\sigma}\right)\right) + (1-p)Q\left(\frac{\eta_i - a_i}{\sigma}\right) \tag{4.16}$$

where $Q(\cdot)$ is the complementary cumulative distribution function of a standard Gaussian distribution defined as

$$Q(x) = \frac{1}{\sqrt{2\pi}} \int_x^{\infty} e^{-\frac{t^2}{2}} dt. \tag{4.17}$$

The probability of a sensor sending a quantized value '1' is given by

$$P(u_i = 1|\theta) = (1-\alpha)Q\left(\frac{\eta_i - a_i}{\sigma}\right) + \alpha\left(p\left(1 - Q\left(\frac{\eta_i - a_i}{\sigma}\right)\right) + (1-p)Q\left(\frac{\eta_i - a_i}{\sigma}\right)\right). \tag{4.18}$$

Proposition 4.1 ([19]) *Under independent attack, the PCRLB is given by* $\mathbf{F}^{-1}$, *where* $\mathbf{F}$ *is the posterior FIM whose elements are given by*

$$F_{11} = \sum_{i=1}^{N} \int_{\theta} p_0(\theta) \frac{(1-2\alpha p)^2 a_i^2 e^{-\frac{(\eta_i - a_i)^2}{\sigma^2}} (x_i - x_t)^2}{2\pi\sigma^2 d_i^4 P_1(1-P_1)} + \frac{1}{\sigma_\theta^2}, \tag{4.19}$$

$$F_{12} = F_{21} = \sum_{i=1}^{N} \int_{\theta} p_0(\theta) \frac{(1-2\alpha p)^2 a_i^2 e^{-\frac{(\eta_i - a_i)^2}{\sigma^2}} (x_i - x_t)(y_i - y_t)}{2\pi\sigma^2 d_i^4 P_1(1-P_1)}, \tag{4.20}$$

$$F_{22} = \sum_{i=1}^{N} \int_{\theta} p_0(\theta) \frac{(1-2\alpha p)^2 a_i^2 e^{-\frac{(\eta_i - a_i)^2}{\sigma^2}} (y_i - y_t)^2}{2\pi\sigma^2 d_i^4 P_1(1-P_1)} + \frac{1}{\sigma_\theta^2}. \tag{4.21}$$

where $P_1 = P(u_i = 1|\theta)$ *is given in* (4.18).

Proof The proof follows from the definition of **F** given in (4.9)–(4.14), and using the following

$$\frac{\partial P_1}{\partial x_t} = -\frac{(1-2\alpha p) a_i e^{-\frac{(\eta_i - a_i)^2}{2\sigma^2}} (x_i - x_t)}{\sigma\sqrt{2\pi} d_i^2}, \tag{4.22}$$

$$\frac{\partial P_1}{\partial y_t} = -\frac{(1-2\alpha p) a_i e^{-\frac{(\eta_i - a_i)^2}{2\sigma^2}} (y_i - y_t)}{\sigma\sqrt{2\pi} d_i^2}. \tag{4.23}$$

□

It can be observed that when $\alpha = 0$, i.e., when all the sensors are honest, the above expression simplifies to the special case of $\mathbf{F_D} = \mathbf{E}_\theta[\mathbf{J}(\theta)]$ where $\mathbf{J}(\theta)$ is the Fisher information matrix derived by Niu et al. in [10] for the case of target localization using quantized data in the absence of Byzantines.

Blinding the Fusion Center

The goal of Byzantines is naturally to cause as much damage to the functionality of the FC as possible. As defined before, this event of causing the maximum possible damage is known as *blinding* the FC which refers to making the FC incapable of using the information contained in the data from the local sensors to estimate the target location. This is clearly the case when the data's contribution to posterior Fisher information matrix, $\mathbf{F_D}$, approaches zero. In this scenario, the best the FC can do is to use the prior to estimate the location. In other words, **F** approaches the prior's contribution to posterior Fisher information, $\mathbf{F_P} = \frac{1}{\sigma_\theta^2}\mathbf{I}$, or the PCRLB approaches $\sigma_\theta^2\mathbf{I}$. Since PCRLB and FIM are matrix-valued and are functions of α, the blinding condition corresponds to the trace of PCRLB tending to $2\sigma_\theta^2$ or the determinant of FIM tending to $\frac{1}{\sigma_\theta^4}$. That is, α^*, the critical power (refer to Chap. 2), is defined as

$$\alpha^* := \min\{\alpha |\, \mathrm{tr}(\mathbf{F}(\alpha)^{-1}) = 2\sigma_\theta^2\}, \tag{4.24}$$

or

$$\alpha^* := \min\left\{\alpha ||\mathbf{F}(\alpha)| = \frac{1}{\sigma_\theta^4}\right\}. \tag{4.25}$$

A closed-form expression for α^* has been derived in [19] by considering the case when all the honest sensors are identical and similarly all the Byzantines are identical. Therefore, in the following, we present the analysis assuming that all the honest sensors use the same local threshold η_H and all the Byzantines use η_B.

Byzantines Modeled as a Binary Symmetric Channel (BSC)

From the FC's perspective, the binary data received from the local sensor is either a false observation of the local sensor with probability $q = \alpha p$ or a true observation with probability $1 - q = (1 - \alpha) + \alpha(1 - p)$. Therefore, the effect of Byzantines, as seen by the FC, can be modeled as a binary symmetric channel (BSC) with transition probability q. It is clear that this virtual 'channel' affects the PCRLB at the FC, which is a function of q. It has been shown in the literature [11] that the Cramér–Rao lower bound (CRLB) of the localization process approaches infinity when this transition probability approaches $\frac{1}{2}$ irrespective of the true location of the target (θ). This result means that the data's contribution to posterior Fisher information $\mathbf{F_D}$ approaches 0 for $q = \alpha p = \frac{1}{2}$. Observe that higher the probability of flipping of the Byzantines (p), the lower the fraction of Byzantines required to blind the FC. So, the minimum fraction of Byzantines, α^*, is $\frac{1}{2}$ corresponding to $p = 1$. In order to blind the network, the Byzantines need to be at least 50% in number and should flip their quantized local observations with probability $p = 1$.

Another interpretation of this problem can be provided from the information theoretical perspective. Since the Byzantines' effect is modeled as a BSC, the capacity of this channel is $C = 1 - H_b(q)$ where $H_b(q)$ is the binary entropy function given by

$$H_b(q) = -q \log_2 q - (1 - q) \log_2 (1 - q). \tag{4.26}$$

The FC receives non-informative data from the sensors or becomes blind, when the capacity approaches 0 which happens when $H_b(q) = 1$ or $q = \frac{1}{2}$. Following the discussion above, we have $\alpha^* = \frac{1}{2}$ and $p = 1$.

It can also be observed that the data's contribution to F_{11} and F_{22} elements of $\mathbf{F}$ given by (4.19) and (4.21) becomes 0 when $\alpha p = \frac{1}{2}$. Once again, we get $\alpha^* = \frac{1}{2}$ and $p = 1$ as the optimal attack strategy to blind the FC. Due to this observation, for the remainder of the analysis, we assume that the Byzantines flip their observations with probability 1, i.e., $p = 1$.

Best Honest and Byzantine Strategies: A Zero-Sum Game

When α, the fraction of Byzantine sensors in the network, is greater than or equal to the critical power α^*, attackers will be able to blind the FC. But when α is not large enough to blind the FC, the Byzantine sensors will try to maximize the damage by making either the trace of $\mathbf{F}^{-1}$, defined as $\mathrm{tr}(\mathbf{F}^{-1})$, as large as possible or the determinant of $\mathbf{F}^{-1}$, defined as $|\mathbf{F}|$, as small as possible. In contrast, the FC will try to minimize $\mathrm{tr}(\mathbf{F}^{-1})$ or maximize $|\mathbf{F}|$. This will result in a game between the FC and each Byzantine attacker where each player has competing goals. Each Byzantine sensor will adjust its threshold η_B to maximize $\mathrm{tr}(\mathbf{F}^{-1})$ or minimize $|\mathbf{F}|$, while the FC will adjust the honest sensor's threshold η_H to minimize $\mathrm{tr}(\mathbf{F}^{-1})$ or maximize $|\mathbf{F}|$. Thus, it is a zero-sum game where the utility of the FC is $-\,\mathrm{tr}(\mathbf{F}^{-1})$ (or $|\mathbf{F}|$) and the utility of the Byzantine sensor is $\mathrm{tr}(\mathbf{F}^{-1})$ (or $-|\mathbf{F}|$) [5]. More formally, consider $\mathrm{tr}(\mathbf{F}^{-1})$ and denote $\mathscr{C}(\eta_H, \eta_B) = \mathrm{tr}(\mathbf{F}^{-1})$ as the cost function adopted by the honest sensors. Let η_H^* and η_B^* denote the best thresholds (strategy) of the honest and the Byzantine sensors, respectively. For a given η_B, η_H^* is computed as

$$\eta_H^* = \arg\min_{\eta_H} \mathscr{C}(\eta_H, \eta_B). \tag{4.27}$$

Similarly, for a given η_H, η_B^* is computed as

$$\eta_B^* = \arg\min_{\eta_B} -\mathscr{C}(\eta_H, \eta_B). \tag{4.28}$$

The solutions to (4.27) and (4.28) characterize the Nash equilibria which are defined as follows [5].

Definition 4.1 A (pure) strategy η_H^* for an honest sensor is a Nash equilibrium (NE) if

$$\mathscr{C}(\eta_H^*, \eta_B^*) \leq \mathscr{C}(\eta_H, \eta_B^*). \tag{4.29}$$

Similarly, a (pure) strategy η_B^* for a Byzantine sensor is a Nash equilibrium (NE) if

$$\mathscr{C}(\eta_H^*, \eta_B^*) \geq \mathscr{C}(\eta_H^*, \eta_B). \tag{4.30}$$

In a zero-sum game, the best strategy for both players[1] is the saddle point, at which none of the players have the incentive to change their strategy. The saddle point for this problem given by (4.27) and (4.28) was found in [19] using traditional methods. First, the set of stationary points of $\mathrm{tr}(\mathbf{F}^{-1})$ defined by $\mathscr{S}$ are identified as

$$\mathscr{S} := \left\{ (\eta_H, \eta_B) | \frac{\partial\, \mathrm{tr}(\mathbf{F}^{-1})}{\partial \eta_H} = \frac{\partial\, \mathrm{tr}(\mathbf{F}^{-1})}{\partial \eta_B} = 0 \right\}. \tag{4.31}$$

The saddle point, (η_H^*, η_B^*), is the one at which the Hessian matrix is indefinite; i.e., the determinant of the Hessian matrix is negative,

$$(\eta_H^*, \eta_B^*) = \left\{ (\eta_H, \eta_B) \in \mathscr{S} | \frac{\partial^2\, \mathrm{tr}(\mathbf{F}^{-1})}{\partial^2 \eta_H} \frac{\partial^2\, \mathrm{tr}(\mathbf{F}^{-1})}{\partial^2 \eta_B} - \left(\frac{\partial^2\, \mathrm{tr}(\mathbf{F}^{-1})}{\partial \eta_H \partial \eta_B} \right)^2 < 0 \right\}. \tag{4.32}$$

Note that the above expressions are based on the use of $\mathrm{tr}(\mathbf{F}^{-1})$. Similar analysis can be carried out when $\mathrm{tr}(\mathbf{F}^{-1})$ is replaced with $|\mathbf{F}|$ as the performance metric.

[1]Here, the set of honest sensors is first player, and the set of Byzantines is the second player.

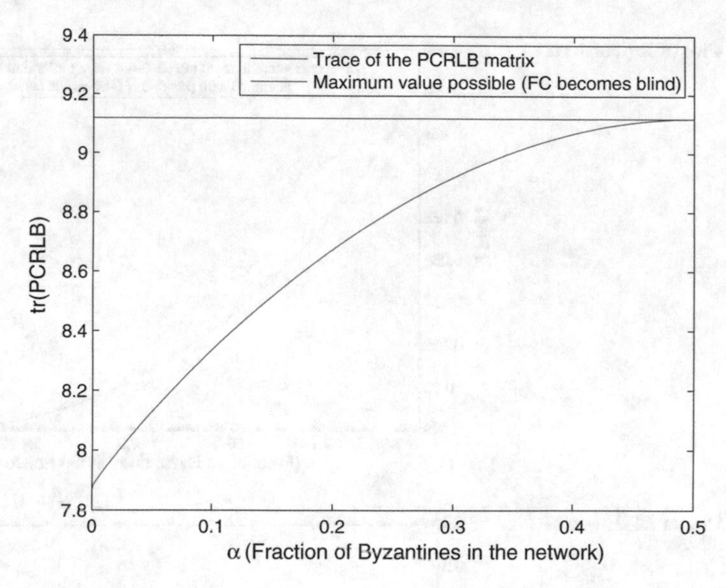

Fig. 4.2 Plot of $\mathrm{tr}(\mathbf{F}^{-1})$ versus α [19]

Numerical Results

In this subsection, simulation results are presented for fundamental limits of Byzantines in a localization problem. Consider the wireless sensor network (WSN) model where $N = 100$ sensors are randomly deployed in a 11×11 square region of interest where the target is located. The target's location is randomly generated from the prior $p_0(\theta)$ with $\sigma_\theta = 2.1352$ such that its 99% confidence region covers the entire ROI. There are $M = \alpha N$ Byzantine sensors present in the network which try to manipulate the data and send falsified information to the FC. Assume that the power at the reference point ($d_0 = 1$) is $P_0 = 200$. The signal amplitude at the local sensor is corrupted by AWGN with standard deviation $\sigma = 3$. In Figs. 4.2 and 4.3, the values of $\mathrm{tr}(\mathbf{F}^{-1})$ and $|\mathbf{F}|$ are plotted against α in the case of an independent attack with $\eta_H = \eta_B = 8.5$. The figures show that when $\alpha = 0.5$, $\mathrm{tr}(\mathbf{F}^{-1})$ approaches $2\sigma_\theta^2$ and $|\mathbf{F}|$ approaches $\frac{1}{\sigma_\theta^4}$. This shows that $\alpha^* = 1/2$; i.e., unless the number of Byzantine sensors is greater or equal to 50 percent of the total number of sensors, the FC cannot be made blind under independent attack. This corroborates the theoretical analysis regarding the PCRLB approaching $\sigma_\theta^2\mathbf{I}$ when α, which is the transition probability of the BSC model, approaches $\frac{1}{2}$. These results can be reproduced for different values of η_H and η_B. Figure 4.4 shows the increase in mean square error (MSE) of the target estimate with α. As the fraction of Byzantines increases, the MSE increases as illustrated in Fig. 4.4. Since the MSE is lower bounded by $\mathrm{tr}(\mathbf{F}^{-1})$, the plot in Fig. 4.4 is always above the plot of $\mathrm{tr}(\mathbf{F}^{-1})$ versus α in Fig. 4.2.

As discussed, when $\alpha < \alpha^*$, there exists a zero-sum game between the FC and Byzantine sensors, in which the optimal strategies are given by the saddle points

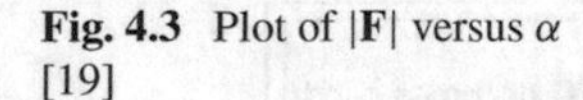

Fig. 4.3 Plot of $|\mathbf{F}|$ versus α [19]

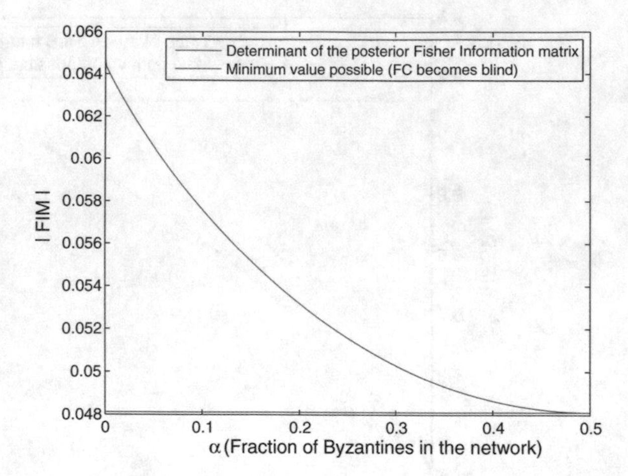

Fig. 4.4 Plot of MSE versus α [19]

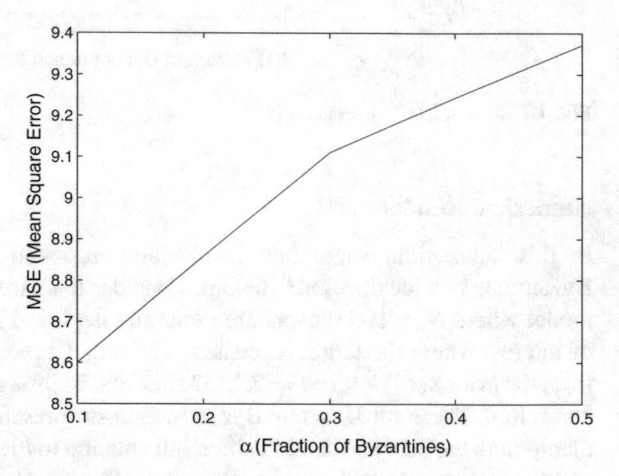

(equilibrium points). In Figs. 4.5 and 4.6, $\mathrm{tr}(\mathbf{F}^{-1})$ is plotted with varying thresholds for $\alpha = 0.4$ and the existence of a saddle point (η_H^*, η_B^*) is observed, which provides optimal strategies for both types of sensors. The saddle point (η_H^*, η_B^*), in this particular example, is at (8.5, 8.5). This result is intuitive as Byzantines flip their decision with probability 1. Therefore, the best strategy for the Byzantines is to use the same best response of the honest sensors and then flip them with probability 1. Similar results can be obtained for different values of α.

Figures 4.7 and 4.8 show similar game-theoretical analysis results for the case of $|\mathbf{F}|$ as the performance metric. The optimal values for this case (η_H^*, η_B^*) are the same (8.5, 8.5). Thus, there exist saddle points (η_H^*, η_B^*) which yield the optimal strategies for both the FC and the Byzantine attackers. We would like to point out that the two objective functions used in the analysis ($\mathrm{tr}(\mathbf{F}^{-1})$ and $|\mathbf{F}|$) need not always result in the same operational point in general. However, in this particular example, this value turns out to be the same, irrespective of the performance metric.

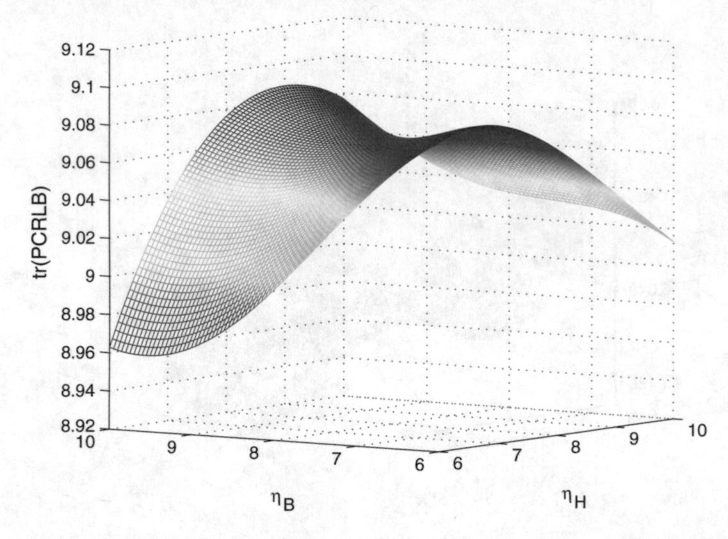

Fig. 4.5 Surface plot of tr($\mathbf{F}^{-1}$) versus honest and Byzantine sensor's threshold, η_H and η_B. The existence of a saddle point is clear [19]

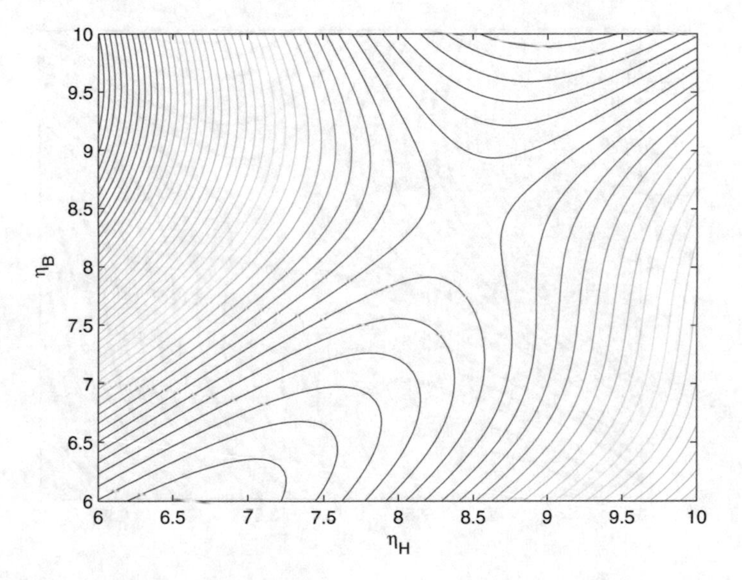

Fig. 4.6 Contour plot of the surface of tr($\mathbf{F}^{-1}$) shown in Fig. 4.5 [19]

Collaborative Attack

Now, consider the case of Byzantine attacks where all the malicious sensors communicate with each other and attack the network in a coordinated fashion. In a collaborative attack, Byzantines collaborate to deteriorate the network's estimation performance and attack the network after colluding with others. Here again, assume

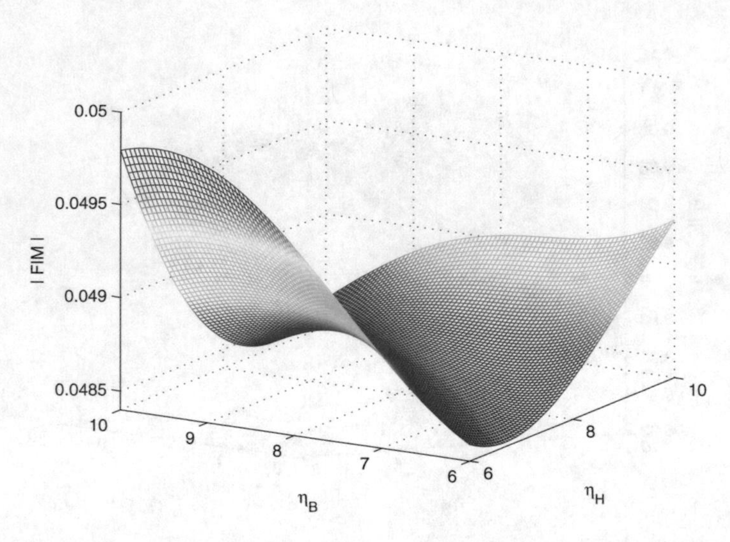

Fig. 4.7 Plot of $|\mathbf{F}|$ versus honest and Byzantine sensor's threshold, η_H and η_B. The existence of a saddle point is clear [19]

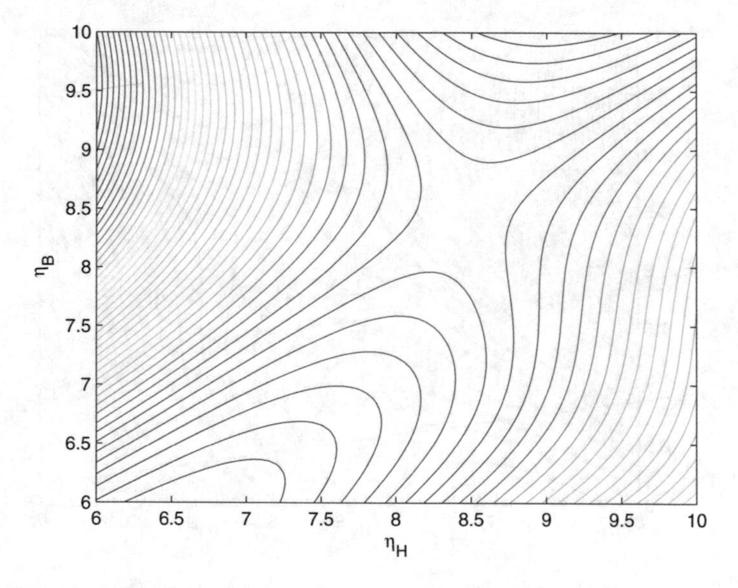

Fig. 4.8 Plot of contour of the surface of $|\mathbf{F}|$ shown in Fig. 4.7 [19]

α to be the fraction of Byzantines present in the network. Analysis of the collaborative attack is significantly more complicated than the independent case. Here, a reasonable lower bound for α^*, namely α_L^*, was derived for this case in [19].

To derive the lower bound for α^*, assume that the exact location of target θ can be perfectly learned by Byzantine sensors due to collaboration. Thus, consider the case where Byzantine attackers know the location of the target and use this information collaboratively to improve their attack on the network. Let us first consider the case where each sensor uses an identical threshold. In such a case, the optimal strategy for the Byzantine sensors will be to send u_i based on the true θ value and their locations. For a given sensor i, its location $\theta_i = [x_i, y_i]$, its observation model, and its threshold $\eta_i = \eta$, the probability of the sensor sending a quantized value 1 as seen by the FC is:

$$P_i(u_i = 1|\theta) = (1-\alpha)P_i^H(u_i = 1|\theta) + \alpha P_i^B(u_i = 1|\theta) \tag{4.33}$$

The Byzantines would like to design their variables α and $P_i^B(u_i = 1|\theta)$ so that the FC becomes blind to the information received from the ith sensor; i.e., the information received from the ith sensor does not help the FC estimate the target location. This can be achieved by making the value $P_i(u_i = 1|\theta)$ a constant value (k) with respect to θ. Let $P_{i,inf}^H := \inf_\theta(P_i^H(u_i = 1|\theta))$ and the $P_{i,sup}^H := \sup_\theta(P_i^H(u_i = 1|\theta))$. Then for the ith sensor, if it was honest and $P_i^H(u_i = 1|\theta) = P_{i,inf}^H$, it would mean that the target is as far away from the ith sensor as possible. However, if the same sensor behaved as a Byzantine, it is reasonable to assume that a sensible Byzantine would send a 1 to the FC with as high a probability as possible, i.e., $P_i^B(u_i = 1|\theta) = 1$. Similarly, if the ith sensor was honest and $P_i^H(u_i = 1|\theta) = P_{i,sup}^H$, then it would mean that the target is as close to the ith sensor as possible and if the same sensor behaved as a Byzantine, under a similar assumption as before, it would send a 1 to the FC with as low a probability as possible, i.e., $P_i^B(u_i = 1|\theta) = 0$. This gives us two equations in two unknowns, $\alpha_L^{*,i}$ and k:

$$(1-\alpha_L^{*,i})P_{i,inf}^H + \alpha_L^{*,i}.1 = k \tag{4.34}$$

and

$$(1-\alpha_L^{*,i})P_{i,sup}^H + \alpha_L^{*,i}.0 = k. \tag{4.35}$$

After solving (4.34) and (4.35),

$$\alpha_L^{*,i} = \frac{P_{i,sup}^H - P_{i,inf}^H}{1 + P_{i,sup}^H - P_{i,inf}^H} \tag{4.36}$$

where $\alpha_L^{*,i}$ is the fraction of malicious sensors required to make sensor i non-informative to the FC. In this case, at the FC,

$$P_i(u_i = 1|\theta) = (1-\alpha_L^{*,i})P_i^H(u_i = 1|\theta) + \alpha_L^{*,i} P_i^B(u_i = 1|\theta) = \frac{P_{i,sup}^H}{1 + P_{i,sup}^H - P_{i,inf}^H} \tag{4.37}$$

which is a constant independent of θ. Thus, when $\alpha \geq \alpha_L^{*,i}$ for a particular honest sensor i, the attackers can have $\alpha - \alpha_L^{*,i}$ fraction of Byzantine sensors act like honest sensors and have the rest $\alpha_L^{*,i}$ Byzantine sensors send 1 with probability as given below:

$$P_i^B(u_i = 1|\theta) = \frac{P_{i,sup}^H - P_i^H(u_i = 1|\theta)}{P_{i,sup}^H - P_{i,inf}^H} \tag{4.38}$$

This causes the FC to become incapable of utilizing the received information from the ith sensor to estimate the target location. In order to guarantee that the FC cannot obtain any useful information from any of its sensors, the minimum required $\alpha_{L,ident}^*$ is given as

$$\alpha_{L,ident}^* = \max_i \alpha_L^{*,i}. \tag{4.39}$$

This provides us with a lower bound, $\alpha_{L,ident}^*$, for the collaborative case under the identical threshold scheme. For the collaborative attack case, it can be observed that $\alpha_L^{*,i}$ given by (4.36) is always ≤ 0.5 which implies that $\alpha_{L,ident}^* \leq 0.5$ which is the α^* obtained in the independent attack case. This shows that if a strategy exists to obtain this lower bound, then the fraction of sensors required to blind the FC would decrease in the collaborative attack case as compared to the independent attack case. A similar observation was made by Rawat et al. in [12] for the primary user detection for cognitive radio networks in the presence of Byzantines.

4.1.3 Analysis from Network Designer's Perspective

So far, the optimal attack strategies were analyzed for the attacker who intends to deteriorate the performance of the localization task at the FC. It was shown that when the fraction of Byzantines in the network is greater than 0.5 and is attacking independently, the FC becomes blind to the data from the sensors and can only estimate the location of the target using the prior information. However, addressing the problem from the network's perspective, one can develop techniques to counter the effect of Byzantines and ensure reliable estimation performance. We now explore such schemes with a specific focus on the independent attack case. Three main schemes are discussed: Byzantine identification scheme [19], design of dynamic nonidentical threshold scheme [19], and coding-theoretical target localization [20]. The Byzantine identification scheme presented herein is similar in principle to the one proposed in [18] for a distributed detection problem. The dynamic nonidentical threshold scheme explores the design of local quantizers by moving away from the traditional static identical thresholds to dynamic nonidentical thresholds that make the system robust to Byzantines. Lastly, the coding-theoretical scheme builds on the DCFECC and DCSD approaches proposed in [21, 22] to develop a computationally more efficient scheme than the traditional maximum likelihood-based schemes. This coding-theoretical scheme is more robust to Byzantine attacks.

4.1.3.1 Byzantine Identification

A basic scheme for the mitigation of independent Byzantine attacks is to identify the Byzantines by observing their behavior over time. Having observed that the optimal for the Byzantines and the honest sensors is to use $\eta_H = \eta_B = \eta$, a scheme has been proposed in [19] to identify the Byzantines present in the network. This scheme is similar to the one proposed in [18] for the cooperative spectrum sensing problem.

The basic idea of this identification scheme is to observe a sensor's behavior over time and decide on whether it behaves closer to an honest or a Byzantine sensor. This is done by comparing the observed values of $\hat{\gamma}_i = P(u_i = 1|\theta)$ to the expected values of $\hat{\gamma}_i^H = P(u_i = 1|\hat{\theta}, i = Honest)$ or $\hat{\gamma}_i^B = P(u_i = 1|\hat{\theta}, i = Byzantine)$. $\hat{\gamma}_i = P(u_i = 1|\theta)$ is estimated in an iterative manner where $\hat{\gamma}_i$ at the $(T+1)$th iteration is calculated as

$$\hat{\gamma}_i(T+1) = \frac{T\hat{\gamma}_i(T) + u_i(T+1)}{T+1}. \tag{4.40}$$

The values of $\hat{\gamma}_i^H$ and $\hat{\gamma}_i^B$ can be calculated using (4.15) and (4.16). It is important to observe here that these values require the location θ which is unknown. These values in this scheme are initialized by using a coarse estimate of the location, $\hat{\theta}$. In order to obtain an initial coarse estimate, $\hat{\theta}$, a procedure similar to the one proposed by Masazade et al. in [8] has been adopted. In this procedure, it is assumed that there are K anchor sensors in the network that have a higher level of security and thereby treated as honest sensors. The initial data is collected at the FC (at time $T = 0$) from these K anchor sensors, and the MMSE estimate is obtained. For the remainder of this scheme, the following model is considered. At every iteration of the algorithm, the sensors send their one-bit data using the pre-designed identical threshold value. Using these N sensors' data of previous T time instants, the FC iteratively updates $\hat{\gamma}_i(T+1)$.

The estimate $\hat{\gamma}_i(T)$ is computed at every iteration T, and a sensor is declared honest or Byzantine based on the test statistic

$$\Lambda_i(T) = \left| \frac{\hat{\gamma}_i(T) - \hat{\gamma}_i^H}{\hat{\gamma}_i(T) - \hat{\gamma}_i^B} \right| \tag{4.41}$$

which is the ratio of the deviations between the estimated behavior of the ith sensor and the expected behavior of an honest sensor, to the estimated behavior of the ith sensor and the expected behavior of a Byzantine sensor. The FC declares a sensor as a Byzantine at time instant T if $\Lambda_i(T)$ is greater than 1. This means that the sensor behaves closer to a Byzantine than an honest sensor. The advantage of this scheme is that it is adaptive such that the sensor's declaration regarding the sensor being honest or Byzantine is based on data from previous time instants. The above formulation (4.41) is purely heuristic, and no optimality has been claimed. It is sub-optimal and easy-to-implement formulation. The rationale behind such a formulation is that in traditional classification/pattern recognition problems, the decision regarding the type (of a sensor) is made by observing the behavior (of the sensor). A sensor is declared as particular type (say Type A) if it behaves closer to the expected behavior of Type A. In this problem, the behavior is characterized by $\hat{\gamma}_i(T)$ and a decision is made by comparing the closeness of this behavior to the expected behavior ($\hat{\gamma}_i{}^H$ or $\hat{\gamma}_i{}^B$). The target location estimation is done after a final decision is made regarding the identity of the Byzantines. The time instant when a final decision is made depends on the particular scenario and is a design criterion. Once, the final decision is made, the data from the sensors identified as Byzantines can be re-flipped and used in the estimation process similar to the adaptive fusion rule designed in [18] for the problem of distributed spectrum sensing in the presence of Byzantines.

Numerical Results

The effectiveness of the identification scheme is presented through numerical results. For this scenario, consider the same network with $N = 100$ sensors uniformly deployed in a 11×11 area as shown in Fig. 4.1. The fraction of the Byzantines is $\alpha = 0.4$; i.e., 40 out of 100 sensors are malicious. It is assumed that there are K anchor sensors as shown in Fig. 4.1. Each sensor measures s_i, the signal amplitude a_i corrupted by AWGN with $\sigma = 3$. The power at the reference distance ($d_0 = 1$) is $P_0 = 200$. The Byzantines and honest sensors use thresholds $\eta_H = \eta_B = 8.5$, which are the optimum thresholds found in Sect. 4.1.2 assuming that the prior distribution of the target location is a normal distribution such that the region of interest (ROI) includes 99% confidence region. Each Byzantine flips its binary observation with probability 1, before sending it to the FC. The results of the proposed Byzantine identification scheme for a specific target location realization can be seen in Figs. 4.9 and 4.10. In Fig. 4.9, the number of wrongly identified sensors (an honest sensor wrongly identified as a Byzantine and vice-versa) is plotted as a function of time for different values of K, the number of anchor sensors. The value of K has been varied to observe the effect of the *coarseness* of the estimate on the Byzantine identification scheme. The graph shows that most of the sensors are correctly identified with time. The number of wrongly identified sensors is maximum for $K = 3$ and minimum for $K = 9$ as expected since the coarse estimate is more accurate when the number of anchor sensors is larger. For $K = 5$, the number of wrongly identified sensors converges to the value of 14 (out of a total of 100 sensors). From this figure, it can be inferred that $K = 5$ is a reasonable number of anchor sensors to be used. In Fig. 4.10, the estimated α given by $\alpha_{est} = \frac{\hat{M}}{N}$, where $\hat{M}$ is the number of sensors identified as Byzantines, is plotted as a function of time for a network with $K = 5$ anchor sensors placed in star formation as shown in Fig. 4.1. As can be seen from Fig. 4.10, α_{est} converges to the value of 0.42. From these figures, it can be inferred that 8 honest sensors (out of 60 honest sensors in the network) have been falsely identified as Byzantines and 6 Byzantines (out of 40 Byzantines present in the network) have been misidentified as honest sensors.

It is important to note that the performance of the proposed scheme depends on the MSE of the initial coarse estimate. Table 4.1 shows the effect of MSE of the estimate

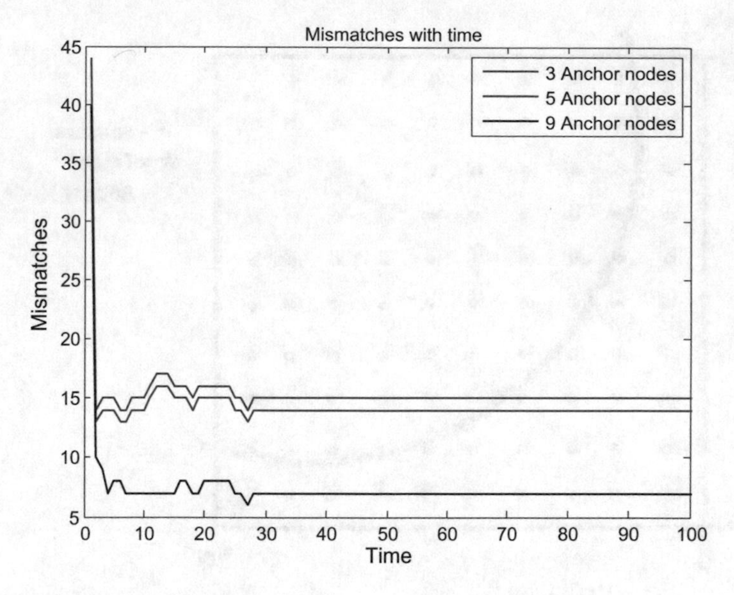

Fig. 4.9 Number of wrongly identified sensors with time [19]

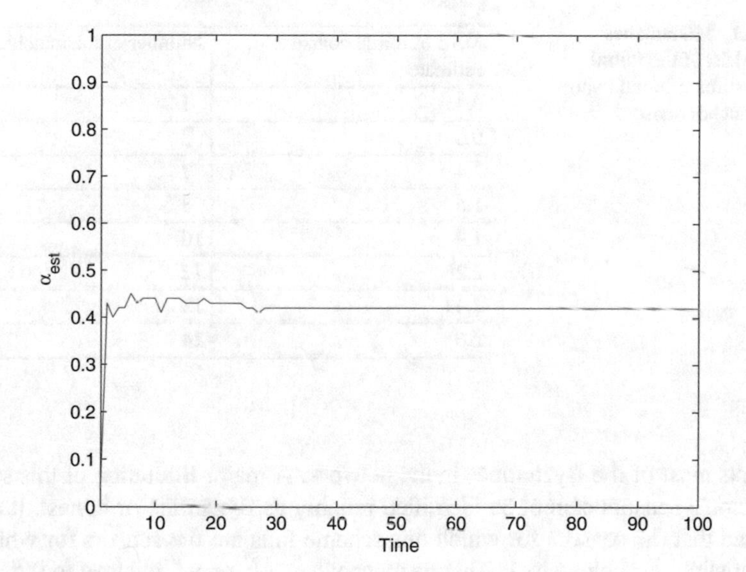

Fig. 4.10 Estimate of α with time [19]

on thc performance of the identification scheme when $K = 5$ anchor sensors are used.

The proposed scheme does not depend on the fraction of Byzantines (α) present in the network, and therefore, the scheme performs well for all possible values of α.

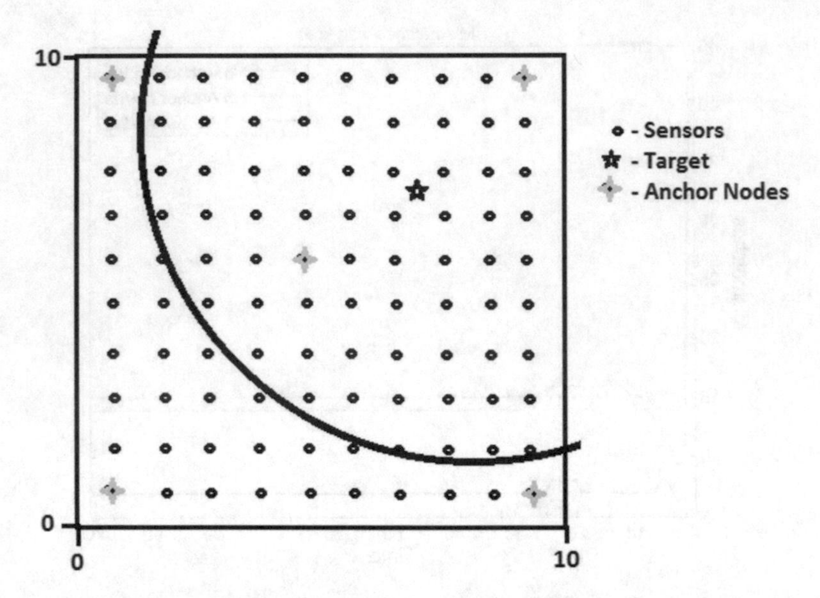

Fig. 4.11 Boundary around which the sensors are 'ambiguous' [19]

Table 4.1 Mismatches versus MSE of the initial coarse estimate found using $K = 5$ anchor sensors

MSE of initial coarse estimate	Number of mismatches
0.4	1
0.6	2
1.4	7
1.6	8
1.9	10
2.28	12
2.44	13
2.6	14

It detects most of the Byzantines in the network. A major limitation of this scheme is that some sensors cannot be identified reliably as Byzantine or honest. It can be observed that the sensors for which this scheme fails are the sensors for which the test statistic Λ_i^T is close to 1. This happens when γ_i^H or γ_i^B is close to 0.5. These are the sensors for which $a_i \approx \eta$ corresponding to some constant $d_i \approx d$. These sensors lie on the boundary region as shown in Fig. 4.11. These *ambiguous* sensors evade the identification process due to the following reason. An ambiguous sensor is defined as the sensor i, for which the quantization threshold is approximately equal to its received amplitude a_i. Therefore, for the Gaussian noise model assumed here, $P(u_i = 1|\hat{\theta}) = \hat{\gamma}_i \approx 0.5$ for an ambiguous sensor irrespective of whether it is an honest sensor or a Byzantine. This implies that these sensors send 1 with approximate probability 0.5. Since the Byzantines flip their decision with probability 1, it cannot be inferred if it was an honest sensor genuinely sending 1 with probability 0.5 or a Byzantine sensor sending 1 with probability 0.5 after flipping its decision. Furthermore, since the received amplitude a_i is constant for ambiguous sensors, they happen to form a circle around the true target location. This *hard* decision regarding the type of sensor results in a high probability of misidentification of the sensors in the region shown in Fig. 4.11. These sensors can therefore be categorized as *ambiguous*. The ratio of the number of these ambiguous sensors on the boundary region to the total number of sensors in a network is expected to be small in practice. In this case, they have negligible impact on estimation performance. However, this number depends on the relative positioning of the target with respect to the local sensors and on the threshold used for quantization at the local sensors. The width of the uncertain zone also depends on sensor noise variance. If this ratio is high, then the estimation performance might be severely degraded. Therefore, a scheme has been designed in [19] where the information from these sensors can be utilized for localization. This problem of boundary sensors can be alleviated as shown in [19] by using nonidentical quantizers, in conjunction with the identification scheme.

Design of Dynamic Nonidentical Quantizers

In this section, the nonidentical quantizer design scheme from [19] is presented to tackle the ambiguity caused by boundary sensors when identical quantizers are used. The estimation model here is sequential and described as follows. At every iteration T, all the N sensors send their one-bit data regarding the location of the target using their local thresholds. At time $T = 0$, the local sensors use the optimal identical thresholds designed for quantization. The FC estimates the target location at every time instant T using Monte Carlo-based MMSE estimation and broadcasts this estimate information to the local sensors. The essential difference between this scheme and the previous one is the feedback between the FC and the local sensors. In other words, according to this scheme, local sensors update their quantizers based on the feedback information (location estimate) they receive from the FC. To design these dynamic quantizers, first consider the case when there are no Byzantines in the network.

Honest Sensors Only

Niu et al. in [10] analyzed the location estimation problem and proposed a threshold design method by minimizing the CRLB on the location estimation error, where the optimal thresholds are found by

$$\min_{\bar{\eta}} V(\bar{\eta}|\theta), \tag{4.42}$$

where $V(\bar{\eta})$ is the trace of the CRLB matrix. In this work, since PCRLB is the performance metric, the optimization problem can be written as

$$\min_{\bar{\eta}} \int V(\bar{\eta}|\theta) p_0(\theta) d\theta. \tag{4.43}$$

This minimization problem is a non-convex vector minimization problem over N variables. This problem can be simplified by making every threshold a function of signal amplitude at the sensor and assuming that all the sensors follow the same functional dependence, i.e.,

$$\eta_i = \eta(a_i), \tag{4.44}$$

where $\eta(\cdot)$ is some function and a_i is the amplitude of the observation at the ith sensor. Since the value of a_i is not known, each threshold is updated iteratively as

$$\eta_i^{T+1} = \eta(\hat{a}_i^T), \tag{4.45}$$

where $\hat{a}_i^T$ is the expected amplitude at the previous time instant which is estimated by using the location estimate of the Tth iteration, $\hat{\theta}^T$. The minimization problem now becomes a variational minimization problem

$$\min_{\eta(\cdot)} \int V(\eta(\cdot)|\theta) p_0(\theta) d\theta. \tag{4.46}$$

This is solved in [19] using a heuristic approach.

Heuristic Approach for Nonidentical Quantizer Design

Note that all the required information about the target location $\theta = [x_t, y_t]$ is completely available in the signal amplitudes a_i's. Therefore, intuitively, if one can accurately estimate a_i from u_i for $i = 1, 2, \ldots, N$, then the target location can be accurately estimated. At any given sensor, this estimation problem is to estimate a using the log-likelihood function given by

$$\ln P(u|a) = (1-u)\ln P(u=0|\eta(\hat{a}), a) + u \ln P(u=1|\eta(\hat{a}), a) \tag{4.47}$$

where a is the signal amplitude at that sensor, $\hat{a}$ is the estimate of the amplitude, and $u \in \{0, 1\}$ is the corresponding quantized bit value.

Proposition 4.2 ([19]) *The posterior Fisher information about the signal amplitude, a, at a given sensor is given by*

$$G[\eta(\cdot)] = \int_A F[\eta(\hat{a}), a] p_A(a) da + \Gamma, \tag{4.48}$$

where $p_A(a)$ is the PDF of the signal amplitude a at the sensor,

$$F[\eta(\hat{a}), a] = \frac{e^{-\frac{(\eta(\hat{a})-a)^2}{\sigma^2}}}{2\pi\sigma^2 [Q(\frac{\eta(\hat{a})-a}{\sigma})][1 - Q(\frac{\eta(\hat{a})-a}{\sigma})]} \tag{4.49}$$

is the data's contribution to posterior FI, and Γ is a constant representing prior's contribution to the posterior FI which is given by

$$\Gamma = -E\left[\frac{d^2 \ln p_A(a)}{da^2}\right]. \tag{4.50}$$

Proof The proof is given in Appendix A.13. □

Now the threshold function $\eta(\cdot)$ can be designed such that it maximizes the posterior Fisher information.

Proposition 4.3 ([19]) *The posterior Fisher information $G[\eta(\cdot)]$ is maximized when the threshold function is $\eta(a) = a$.*

Proof Since the second term on the right-hand side of (4.48) is not a function of η, consider only the first term. The maximization of the first term with respect to $\eta(\cdot)$ is a functional maximization problem, and it can be solved using the Euler–Lagrange equation from variational calculus [17] given by

$$\frac{\partial F}{\partial \eta} = \frac{d}{da}\frac{\partial F}{\partial \eta^{(1)}}, \tag{4.51}$$

where $\eta^{(1)}$ is the first derivative of η with respect to a. Since $F[\eta(\hat{a}), a]$ is independent of $\eta^{(1)}$, the Euler–Lagrange equation reduces to $\frac{\partial F}{\partial \eta} = 0$ or $F =$ constant with respect to η. From (4.49), this gives the result that $\eta(a) = a$ for F to be a constant with respect to η.[1]

Such an analysis was also carried out numerically by Ribeiro et al. in [13], where the authors plotted the CRLB against $(\eta - a)$ to show that $\eta(a) = a$ minimizes the CRLB or in other words maximizes the Fisher information. Therefore, the thresholds are designed such that, $\eta_i^{T+1} = \eta(\hat{a}_i^T) = \hat{a}_i^T$ which means that the threshold of the ith sensor at time $(T + 1)$ is the estimated amplitude at this sensor at the previous time instant T. This amplitude is estimated by using the previous time instant's location estimate, $\hat{\theta}^T$, which is broadcast by the FC to the local sensors. It is important to note that this result is expected as such a threshold design will ideally yield the maximum entropy as it results in $P(u_i = 1|\theta) = P(u_i = 0|\theta) = \frac{1}{2}$.

Game Between Honest and Byzantine Sensors

The situation changes when there are honest sensors as well as Byzantine sensors present in the system. Since the Byzantines' aim is to deteriorate the system performance, they do not necessarily use the threshold design specified by the FC. Instead, Byzantines use their own threshold function $\eta^B(\cdot)$ and they flip their decisions with probability p. Let the threshold function of the honest sensors be $\eta^H(\cdot)$. The following proposition characterizes the posterior Fisher information as a function of the threshold functions.

---

[1] Interested reader is referred to Sect. 2.2 of [17] for further information on Euler–Lagrange equation and variational calculus.

Proposition 4.4 ([19]) *The posterior Fisher information about the signal amplitude, a, at a given sensor in the presence of Byzantines is given by*

$$G[\eta^H(\cdot),\eta^B(\cdot)] = \int_A F[\eta^H(.),\eta^B(.),a]p_A(a)da + \Gamma, \tag{4.52}$$

where the constant Γ is given in (4.50)*, $p_A(a)$ is the PDF of the signal amplitude a at the sensor, and*

$$F[\eta^H(\cdot),\eta^B(\cdot),a] = \frac{\left(-\alpha(2p-1)e^{-\frac{(\eta^B(\hat{a})-a)^2}{2\sigma^2}} + (1-\alpha)e^{-\frac{(\eta^H(\hat{a})-a)^2}{2\sigma^2}}\right)^2}{2\pi\sigma^2[P_1][1-P_1]}, \tag{4.53}$$

with P_1 is defined as the probability of the sensor sending 1 *as seen by the FC*

$$P_1 = \alpha\left(p\left(1-Q\left(\frac{\eta^B(\hat{a})-a}{\sigma}\right)\right) + (1-p)Q\left(\frac{\eta^B(\hat{a})-a}{\sigma}\right)\right) + (1-\alpha)Q\left(\frac{\eta^H(\hat{a})-a}{\sigma}\right). \tag{4.54}$$

Proof The proof is given in Appendix A.14. □

In this case when both honest sensors and Byzantines are present, the problem can again be modeled as a zero-sum game where the objective of the FC is to maximize the posterior Fisher information $G[\eta^H(.),\eta^B(.)]$ whereas the objective of the Byzantine sensor is to minimize $G[\eta^H(.),\eta^B(.)]$. This problem can be solved by examining the expression of G in (4.52). Under the scenario that each sensor behaves independently, it can be shown that the Fisher information (FI) given by (4.53) is maximized when honest sensors set their thresholds as $\eta^H(\hat{a}) = \hat{a}$ regardless of the value of η^B. For Byzantines, there are two cases to be considered: $\alpha p < \frac{1}{2}$ and $\alpha p \geq \frac{1}{2}$. For $\alpha p < \frac{1}{2}$, the Byzantines, who try to minimize this posterior FI, achieve minimization similarly by setting $\eta^B(\hat{a}) = \hat{a}$ regardless of the value of η^H. This result is expected as the Byzantines flip their observations with a probability p. It is important to observe that when $\eta^B(\hat{a}) = \eta^H(\hat{a}) = \hat{a}$, it implies that the honest sensors send 0/1 with probability approximately equal to $\frac{1}{2}$. Also, observe that if the Byzantines also use this thresholding scheme, the probability of a Byzantine sending a 1 is

$$\begin{aligned} P(u=1|a, Byzantine) = p\left(1-Q\left(\frac{\eta^B(\hat{a})-a}{\sigma}\right)\right) \\ + (1-p)Q\left(\frac{\eta^B(\hat{a})-a}{\sigma}\right), \end{aligned} \tag{4.55}$$

which becomes $\frac{1}{2}$, when $\eta^B(\hat{a}) = \hat{a}$, irrespective of the value of p. Similar to the honest sensors, the Byzantines maximize the entropy as well and eventually benefit the network in the localization task as discussed earlier. This result can also be interpreted by using the BSC modeling of Byzantines as done in [19]. In this model, the honest sensors try to transmit the data (0/1) such that the capacity is achieved. The capacity is achieved when the input follows uniform distribution. The proposed threshold scheme makes $P(u_i = 0/1|\theta) = \frac{1}{2}$, thereby achieving capacity of this 'Byzantine' BSC. For $\alpha p \geq \frac{1}{2}$, the Byzantines need to use $\eta^B(\hat{a}) = \pm\infty$, that is, always send a 0 or 1. However, in this case, the network would again eventually overcome the actions of Byzantines as the FC can easily identify the Byzantines using the Byzantine identification scheme previously proposed.

It is worthwhile to point out here that any non-honest strategy used by the Byzantines can be identified by the FC and therefore the effect of Byzantines can be mitigated. Therefore, utilizing dynamic nonidentical quantizers ensures that the Byzantines become *ineffective* in their attack strategy and the network eventually mitigates the actions of Byzantines. It is also important to observe that this particular framework ensures that the network is robust for any fraction (α) of the Byzantines in the system. The trade-off is that the FC needs to broadcast the target location estimate at each iteration which increases the system complexity and consumes more system resources compared to using static identical quantizers. Static thresholds are set only once in the beginning, and they stay constant throughout the estimation process. In the identification procedure, the FC tries to identify sensors based on their observed behavior over time, which requires each sensor to send their decisions to the FC in a continuous manner. In contrast, dynamic thresholds are adjusted dynamically at each sensor using the feedback from the FC. As one would expect, feedback not only improves estimation performance but it also makes the network more robust to Byzantine attacks.

Numerical Results

The superiority of the dynamic nonidentical quantizer design over the static identical quantizer design was presented via simulations in [19]. For this scenario, consider a network with $N = 100$ sensors uniformly deployed in a 11×11 area same as before (refer to Fig. 4.1). The power at the reference distance ($d_0 = 1$) is again set as $P_0 = 200$. This setup yielded an optimal identical threshold as $\eta = 8.5$. During the first iteration, the location $\theta^{(1)}$ is estimated using identical thresholds. After obtaining $\hat{\theta}^{(1)}$, the estimates in the next iterations are calculated by using the proposed dynamic nonidentical threshold design scheme as $\eta_i^{T+1} = \hat{a}_i^T$. MMSE estimators are implemented using the important sampling method with $N_p = 10000$ particles. Figure 4.12 shows the mean square error (MSE) values of the estimators for 1000 Monte Carlo realizations of θ compared to the MSE of those using identical thresholds. As can be seen from Fig. 4.12, the estimation error reduces significantly (by around 70% in 3 iterations) when nonidentical dynamic threshold quantizers are used as compared to the identical threshold quantizers. This motivates the honest sensors to use the designed nonidentical thresholds. As discussed above, the Byzantines are ineffective in this scheme.

When each sensor uses this dynamic nonidentical quantizer design scheme for the case of collaborative attack mentioned before, the analysis is extremely difficult; however, the following was conjectured in [19]. The largest deviation in the current estimate caused by the Byzantines is limited to the confidence interval of the previous

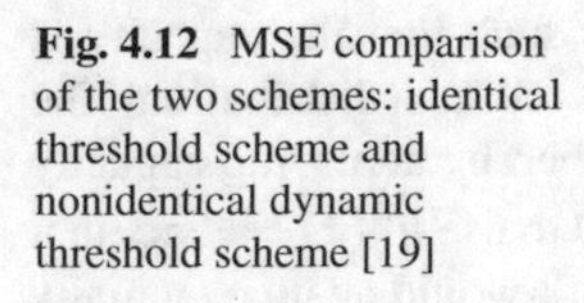
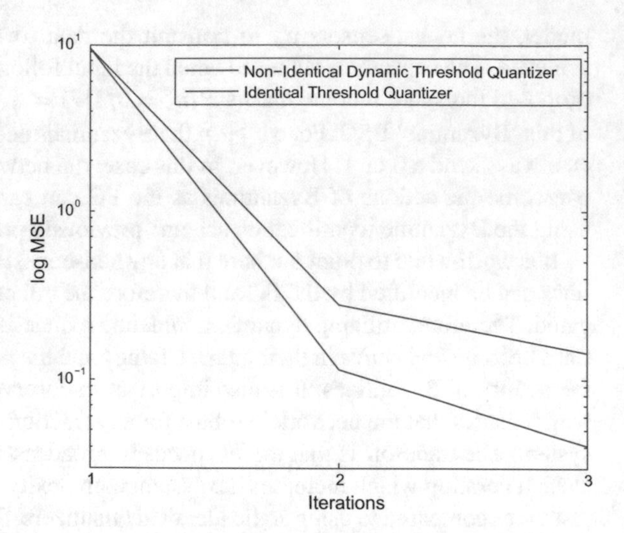

Fig. 4.12 MSE comparison of the two schemes: identical threshold scheme and nonidentical dynamic threshold scheme [19]

estimate since, otherwise, they would be easily identified as outliers at the FC. This limits the attacking power of each Byzantine, and hence, more number of Byzantines are required to cause the same blinding effect at the FC. Therefore, it was conjectured that $\alpha^*_{L,non-ident} \geq \alpha^*_{L,ident}$, where $\alpha^*_{L,non-ident}$ denotes the fraction of Byzantines required to blind the FC under dynamic nonidentical quantizer scheme.

So far, two schemes have been presented to mitigate the effect of Byzantines in the network. While the first scheme is passive and only learns the Byzantines' identity over time, the second scheme changes the system and uses a dynamic nonidentical threshold scheme. However, both the schemes still use maximum likelihood or MMSE-based optimal estimators. Another scheme has been proposed in the literature [20] where the estimation/localization approach has been modified to a sub-optimal but easy-to-implement approach. This scheme works on the idea that classification is easier than estimation and therefore formulates the localization problem as hierarchical classification using error-correcting codes. This scheme is described in the following, and its robustness to Byzantines is presented.

Error-Correcting Code-Based Localization

It is universally understood that a typical estimation problem is computationally more expensive than a typical classification problem. This is intuitively expected since classification is a search problem among finite and typically low number of choices while an estimation problem is an open-ended problem with infinite choices within a bounded region. Based on this intuition, Vempaty et al. proposed a localization scheme in [20] that is computationally less expensive than traditional approaches and also robust to Byzantines in the network.

Consider the setup discussed in this chapter, where N sensors are deployed in a network and a target at an unknown location in the network emitting power. The algorithm is iterative in which at every iteration, an M-ary hypothesis test is performed at the FC and accordingly the region of interest (ROI) is split into M regions. The FC, through feedback, declares the selected region as the ROI for the next iteration. The M-ary hypothesis test solves a classification problem where each sensor sends binary quantized data based on an approach proposed by [21].

In this classification approach, called distributed classification fusion using error-correcting codes (DCFECC), a code matrix C is selected to perform both local decision and fault-tolerant fusion at the FC. The code matrix is an $M \times N$ matrix with elements $c_{(j+1)i} \in \{0, 1\}$, $j = 0, 1, \ldots, M-1$ and $i = 1, \ldots, N$. Each hypothesis H_j is associated with a row in the code matrix C, and each column represents a binary decision rule at the local sensor. The optimal code matrix is designed off-line using techniques such as simulated annealing or cyclic column replacement [21]. Each local sensor employs the decision rule corresponding to its column in C and transmits its decision to FC. After receiving the binary decisions $\boldsymbol{u} = [u_1, \ldots, u_N]$ from the local sensors, the FC performs minimum Hamming distance-based fusion [21] and decides on the hypothesis H_j for which the Hamming distance between the row of C corresponding to H_j for $j = 0, \ldots, M-1$ and the received vector $\boldsymbol{u}$ is minimum. In other words, the fusion rule used by the fusion center is to decide H_j, where

$$j = \arg\min_{0 \le j \le M-1} d_H(\boldsymbol{u}, \mathbf{r}_j), \tag{4.56}$$

$d_H(\mathbf{x}, \mathbf{y})$ is the Hamming distance between vectors $\mathbf{x}$ and $\mathbf{y}$, and $\mathbf{r}_j$ is the row of C corresponding to hypothesis H_j. The tie-breaking rule is to randomly pick a code word from those with the same smallest Hamming distance to the received vector. Due to the minimum Hamming distance-based fusion scheme, the DCFECC approach can also handle missing data. When a sensor does not return any decision, its contribution to the Hamming distance between the received vector and every row of the code matrix is the same and therefore the row corresponding to the minimum Hamming distance remains unchanged. The error-correcting property of the code matrix provides fault tolerance capability [21]. It is important to note that this DCFECC scheme is under the assumption that $N > M$ and the performance of the scheme depends on the minimum Hamming distance d_{min} of the code matrix C.

For the localization problem, the DCFECC approach was incorporated at every iteration in [20]. The code matrix is of size $M \times N$ with elements $c_{(j+1)i} \in \{0, 1\}$, $j = 0, 1, \cdots, M-1$ and $i = 1, \cdots, N$, where each row here represents a possible region and each column i represents ith sensor's binary decision rule. After receiving the binary decisions $\boldsymbol{u} = [u_1, u_2, \cdots, u_N]$ from local sensors, the FC performs minimum Hamming distance-based fusion. In this way, the search space for target location is reduced at every iteration and the search is stopped based on a predetermined stopping criterion. The optimal splitting of the ROI at every iteration depends on the topology of the network and the distribution of sensors in the network. For a given network topology, the optimal region split can be determined off-line using k-means clustering [2] which yields Voronoi regions [1] containing equal number of sensors in every region. For instance, when the sensors are deployed in a uniform grid, the optimal splitting is uniform as shown in Fig. 4.13. For deriving analytical results, a symmetric

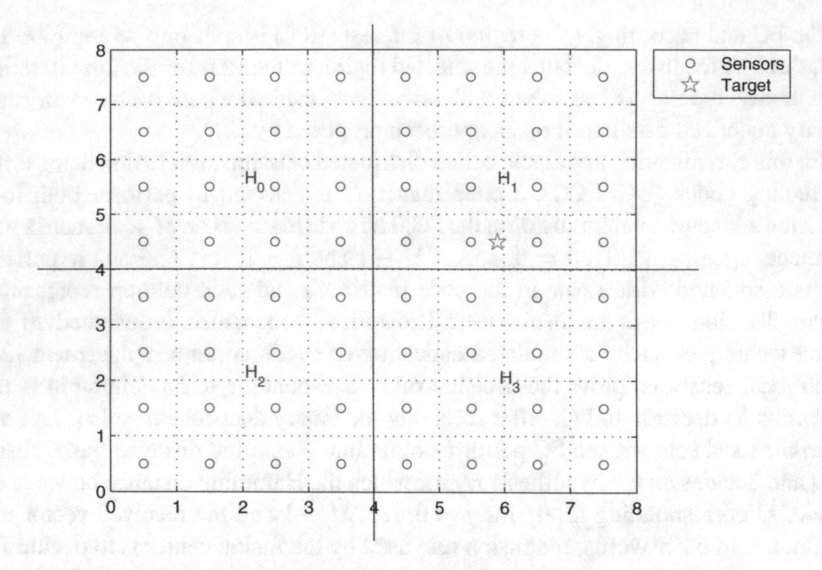

Fig. 4.13 Equal region splitting of the ROI for localization via a four-hypothesis test

sensor deployment, such as a grid, was considered in [20]. Such a deployment results in a one-to-one correspondence between sensors across regions which was leveraged in the derivations.

To understand the details of the scheme and conduct basic analysis, consider the ideal case where all the sensors are benign and the channels between the local sensors and the FC are perfect. Therefore, the binary decisions received at the FC are the same as the binary decisions made by the local sensors, i.e., $u_i = D_i$, for $i = 1, \cdots, N$ where D_i is defined in (4.3). After introducing the basic scheme, the extended scheme that handles unreliable data sources such as Byzantines to ensure secure estimation is presented. As before, the FC estimates the target location using the received data $\boldsymbol{u}$.

Since there are N sensors which are split into M regions, the number of sensors in the new ROI after every iteration is reduced by a factor of M. After k iterations, the number of sensors in the ROI is $\frac{N}{M^k}$ and therefore the code matrix at the $(k+1)$th iteration would be of size $M \times \frac{N}{M^k}$.[2] Since the code matrix should always have more columns than rows, $k^{stop} < \log_M N$, where k^{stop} is the number of iterations after which the scheme terminates. After k^{stop} iterations, there are only $\frac{N}{M^{k^{stop}}}$ sensors present in the ROI and a coarse estimate $\hat{\theta} = [\hat{\theta}_x, \hat{\theta}_y]$ of the target's location can be obtained by taking an average of locations of the $\frac{N}{M^{k^{stop}}}$ sensors present in the ROI:

[2]It is assumed that N is divisible by M^k for $k = 0, 1, \ldots, \log_M N - 1$.

$$\hat{\theta}_x = \frac{M^{k^{stop}}}{N} \sum_{i \in ROI_{k^{stop}}} x_i \tag{4.57}$$

$$\text{and} \quad \hat{\theta}_y = \frac{M^{k^{stop}}}{N} \sum_{i \in ROI_{k^{stop}}} y_i, \tag{4.58}$$

where $ROI_{k^{stop}}$ is the ROI at the last step.

Since the scheme is iterative, the code matrix needs to be designed at every iteration. Observing the structure of the problem, the code matrix can be designed in a simple and efficient way as described below. As pointed out before, the size of the code matrix C^k at the $(k+1)$th iteration is $M \times \frac{N}{M^k}$, where $0 \leq k \leq k^{stop}$. Each row of this code matrix C^k represents a possible hypothesis described by a region in the ROI. Let R_j^k denote the region represented by the hypothesis H_j for $j = 0, 1, \cdots, M-1$, and let S_j^k represent the set of sensors that lie in the region R_j^k. Also, for every sensor i, there is a unique corresponding region in which the sensor lies and the hypothesis of the region is represented as $r^k(i)$. It is easy to see that $S_j^k = \{i \in ROI_k | r^k(i) = j\}$. The code matrix is designed in such a way that for the jth row, only those sensors that are in R_j^k have a '1' as their elements in the code matrix. In other words, the elements of the code matrix are given by

$$c_{(j+1)i}^k = \begin{cases} 1 & \text{if } i \in S_j^k \\ 0 & \text{otherwise} \end{cases}, \tag{4.59}$$

for $j = 0, 1, \cdots, M-1$ and $i \in ROI_k$.

The above construction can also be viewed as each sensor i using a threshold η_i^k for quantization (as described in (4.3)). Let each region R_j^k correspond to a location θ_j^k for $j = 0, 1, \cdots, M-1$, which in our case is the center of the region R_j^k. Each sensor i decides on a '1' if and only if the target lies in the region $R_{r^k(i)}^k$. Every sensor i, therefore, performs a binary hypothesis test described as follows:

$$\begin{aligned} H_1 &: \theta^k \in R_{r^k(i)}^k \\ H_0 &: \theta^k \notin R_{r^k(i)}^k. \end{aligned} \tag{4.60}$$

If d_{i,θ_j^k} represents the Euclidean distance between the ith sensor and θ_j^k for $i = 1, 2, \cdots, N$ and $j = 0, 1, \cdots, M-1$, then $r^k(i) = \arg\min_l d_{i,\theta_l^k}$. Therefore, the condition $\theta^k \in R_{r^k(i)}^k$ can be abstracted as a threshold η_i^k on the local sensor signal amplitude given by

$$\eta_i^k = \frac{\sqrt{P_0}}{d_{i,\theta_{r^k(i)}^k}}. \tag{4.61}$$

This ensures that if the signal amplitude at the ith sensor is above the threshold η_i^k, then θ^k lies in region $R^k_{r^k(i)}$ leading to minimum distance decoding.

Performance Analysis

The performance of this localization scheme is now analyzed. An analytically tractable metric to analyze the performance of the proposed scheme is the probability of detection of the target region. It is an important metric when the final goal of the target localization task is to find the approximate region or neighborhood where the target lies rather than the true location itself. Since the final ROI could be one of the M regions, a metric of interest is the probability of 'zooming' into the correct region. In other words, it is the probability that the true location and the estimated location lie in the same region.

The final region of the estimated target location is the same as the true target location if and only if we 'zoom' into the correct region at every iteration of the proposed scheme. If P_d^k denotes the detection probability at the $(k+1)$th iteration, the overall detection probability is given by

$$P_D = \prod_{k=0}^{k^{stop}} P_d^k. \tag{4.62}$$

Exact Analysis

Consider the $(k+1)$th iteration, and define the received vector at the FC as $\boldsymbol{u}^k = [u_1^k, u_2^k, \cdots, u_{N_k}^k]$, where N_k are the number of local sensors reporting their data to FC at $(k+1)$th iteration. Let $\mathscr{D}_j^k$ be the decision region of jth hypothesis defined as follows:

$$\mathscr{D}_j^k = \{\boldsymbol{u}^k | d_H(\boldsymbol{u}^k, \boldsymbol{c}_{j+1}^k) \leq d_H(\boldsymbol{u}^k, \boldsymbol{c}_{l+1}^k) \text{ for } 0 \leq l \leq M-1\}, \tag{4.63}$$

where $d_H(\cdot, \cdot)$ is the Hamming distance between two vectors and $\boldsymbol{c}_{j+1}^k$ is the code word corresponding to hypothesis j in code matrix C^k. Then, define the reward $r_{\boldsymbol{u}^k}^{j,k}$ associated with the hypothesis j as

$$r_{\boldsymbol{u}^k}^{j,k} = \begin{cases} \frac{1}{q_{\boldsymbol{u}^k}} & \text{when } \boldsymbol{u}^k \in \mathscr{D}_j^k \\ 0 & \text{otherwise} \end{cases}, \tag{4.64}$$

where $q_{\boldsymbol{u}^k}$ is the number of decision regions to whom $\boldsymbol{u}^k$ belongs to. Note that $q_{\boldsymbol{u}^k}$ can be greater than one when there is a tie at the FC. Since the tie-breaking rule is to choose one of them randomly, the reward is given by (4.64). According to (4.64), the detection probability at the $(k+1)$th iteration is given by

$$
\begin{aligned}
P_d^k &= \sum_{j=0}^{M-1} P(H_j^k) \sum_{\boldsymbol{u}^k \in \{0,1\}^{N_k}} P(\boldsymbol{u}^k | H_j^k) r_{\boldsymbol{u}^k}^{j,k} \\
&= \frac{1}{M} \sum_{j=0}^{M-1} \sum_{\boldsymbol{u}^k \in \mathscr{D}_j^k} \left(\prod_{i=1}^{N_k} P(u_i^k | H_j^k) \right) \frac{1}{q_{\boldsymbol{u}^k}},
\end{aligned} \tag{4.65}
$$

where $P(u_i^k|H_j^k)$ denotes the probability that the sensor i sends the bit $u_i^k \in \{0, 1\}$, $i = 1, 2, \cdots, N_k$, when the true target is in the region R_j^k corresponding to H_j^k at the $(k+1)$th iteration.

From the system model described earlier,

$$
P(u_i^k = 1 | H_j^k) = E_{\theta|H_j^k} \left[P(u_i^k = 1 | \theta, H_j^k) \right]. \tag{4.66}
$$

Since (4.66) is complicated, it can be approximated using θ_j^k which is the center of the region R_j^k. (4.66) now simplifies to

$$
P(u_i^k = 1 | H_j^k) \approx Q\left(\frac{\eta_i^k - a_{ij}^k}{\sigma} \right), \tag{4.67}
$$

where η_i^k is the threshold used by the ith sensor at kth iteration, σ^2 is the noise variance, a_{ij}^k is the signal amplitude received at the ith sensor when the target is at θ_j^k, and $Q(x)$ is the complementary cumulative distribution function of standard Gaussian and is given by

$$
Q(x) = \frac{1}{\sqrt{2\pi}} \int_x^{\infty} e^{(-t^2/2)} dt. \tag{4.68}
$$

Using (4.62), the probability of detection of the target region is the product of detection probabilities at every iteration k. It is clear from the derived expressions that the exact analysis of the detection probability is complicated and therefore some analytical bounds have been derived on the performance of the scheme in [20].

Performance Bounds

For performance analysis, a lemma in [23] was used, which is stated here for the sake of completeness.

Lemma 4.1 ([23]) *Let* $\{Z_j\}_{j=1}^{\infty}$ *be independent antipodal random variables with* $Pr[Z_j = 1] = q_j$ *and* $Pr[Z_j = -1] = 1 - q_j$. *If* $\lambda_m \triangleq E[Z_1 + \cdots + Z_m]/m < 0$, *then*

$$
Pr\{Z_1 + \cdots + Z_m \geq 0\} \leq (1 - \lambda_m^2)^{m/2}. \tag{4.69}
$$

Using this lemma, the performance bounds on the scheme have been derived.

Lemma 4.2 ([20]) *Let $\theta \in R_j^k$ be the fixed target location. Let $P_e^k(\theta)$ be the error probability of detection of the target region given θ at the $(k+1)$th iteration. For the received vector of $N_k = N/M^k$ observations at the $(k+1)$th iteration, $\mathbf{u}^k = [u_1^k, \cdots, u_{N_k}^k]$, assume that for every $0 \leq j, l \leq M-1$ and $l \neq j$,*

$$\sum_{i \in S_j^k \cup S_l^k} q_{i,j}^k < \frac{N_k}{M} = \frac{N}{M^{k+1}}, \tag{4.70}$$

where $q_{i,j}^k = P\{z_{i,j}^k = 1|\theta\}$, $z_{i,j}^k = 2(u_i^k \oplus c_{(j+1)i}^k) - 1$, and $C^k = \{c_{(j+1)i}^k\}$ is the code matrix used at the $(k+1)$th iteration. Then,

$$P_e^k(\theta) \leq \sum_{0 \leq l \leq M-1, l \neq j} \left(1 - \frac{\left(\sum_{i \in S_j^k \cup S_l^k}(2q_{i,j}^k - 1)\right)^2}{d_{m,k}^2}\right)^{d_{m,k}/2} \tag{4.71}$$

$$\leq (M-1)\left(1 - \left(\lambda_{j,max}^k(\theta)\right)^2\right)^{d_{m,k}/2}, \tag{4.72}$$

where $d_{m,k}$ is the minimum Hamming distance of the code matrix C^k given by $d_{m,k} = \frac{2N}{M^{k+1}}$ due to the structure of our code matrix and

$$\lambda_{j,max}^k(\theta) \triangleq \max_{0 \leq l \leq M-1, l \neq j} \frac{1}{d_{m,k}} \sum_{i \in S_j^k \cup S_l^k} (2q_{i,j}^k - 1). \tag{4.73}$$

Proof The proof is provided in Appendix A.15. □

The probabilities $q_{i,j}^k = P\{u_i^k \neq c_{(j+1)i}^k|\theta\}$ can be easily computed as given below. For $0 \leq j \leq M-1$ and $1 \leq i \leq N_k$, if $i \in S_j^k$,

$$\begin{aligned} q_{i,j}^k &= P\{u_i^k = 0|\theta\} \\ &= 1 - Q\left(\frac{(\eta_i^k - a_i)}{\sigma}\right), \end{aligned} \tag{4.74}$$

where η_i^k is the threshold used by the ith sensor at $(k+1)$th iteration, σ^2 is the noise variance, and a_i is the amplitude received at the ith sensor given by (4.1) when the target is at θ. If $i \notin S_j^k$, $q_{i,j}^k = 1 - P\{u_i^k = 0|\theta\}$.

For ease of analysis, the following assumption has been made that will be used in the main theorem that presents the optimality of the scheme. Note that, the scheme can still be applied to those WSNs where the assumption does not hold.

Assumption 4.1 For any target location $\theta \in R_j^k$ and any $0 \leq k \leq k^{stop}$, there exists a bijection function f from S_j^k to S_l^k, where $0 \leq l \leq M-1$ and $l \neq j$, such that

$$f(i_j) = i_l,$$

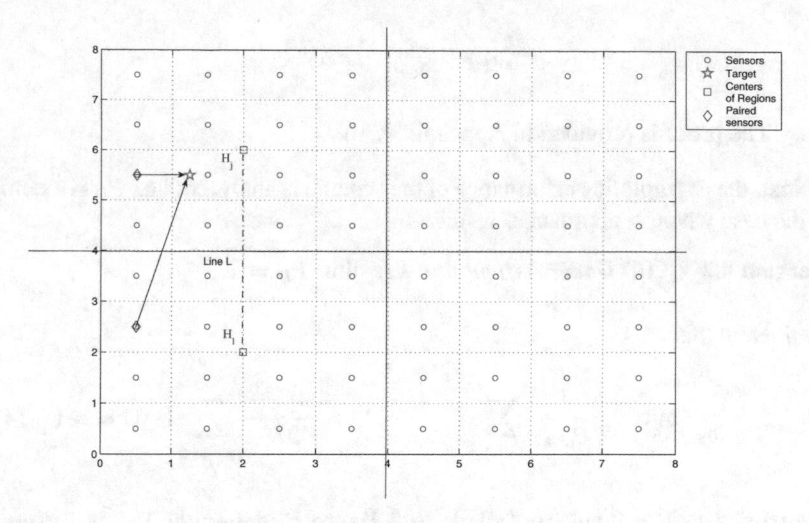

Fig. 4.14 ROI with an example set of paired sensors

$$\eta^k_{i_j} = \eta^k_{i_l},$$

and

$$d_{i_j} < d_{i_l},$$

where $i_j \in S^k_j$, $i_l \in S^k_l$, and d_{i_j} (d_{i_l}) is the distance between θ and sensor i_j (i_l).

One example of WSNs that satisfies this assumption is given in Fig. 4.14. For every sensor $i_j \in S^k_j$, due to the symmetric region splitting, there exists a corresponding sensor $i_l \in S^k_l$ which is symmetrically located as described in the following: Join the centers of the two regions, and draw a perpendicular bisector to this line as shown in Fig. 4.14. The sensor $i_l \in S^k_l$ is the sensor located symmetrically to sensor i_j on the other side of the line L. These are the sensors for which the thresholds are the same. In other words, due to the symmetric placement of the sensors, $\eta^k_{i_j} = \eta^k_{i_l}$. Clearly, when $\theta \in R^k_j$, $d_{i_j} < d_{i_l}$.

Theorem 4.1 ([20]) *Let P_D be the probability of detection of the target region given by* (4.62), *where P^k_d is the detection probability at the* $(k+1)$*th iteration. Under Assumption 4.1,*

$$P^k_d \geq 1 - (M-1)\left(1 - (\lambda^k_{max})^2\right)^{d_{m,k}/2}, \tag{4.75}$$

where

$$\lambda^k_{max} \triangleq \max_{0 \leq j \leq M-1} \lambda^k_{j,max}$$

and

$$\lambda_{j,max}^{k} \triangleq \max_{\theta \in R_j^k} \lambda_{j,max}^{k}(\theta).$$

Proof The proof is provided in Appendix A.16. □

Next, the asymptotic performance of the scheme is analyzed; i.e., P_D is examined for the case when N approaches infinity.

Theorem 4.2 ([20]) *Under Assumption 4.1,* $\lim_{N\to\infty} P_D = 1$.

Proof Note that

$$\lambda_{j,\max}^{k} = \max_{0\le l\le M-1, l\ne j} \frac{1}{d_{m,k}} \sum_{i\in S_j^k \cup S_l^k} (2q_{i,j}^k - 1) > \frac{M^{k+1}}{2N} \sum_{i\in S_j^k \cup S_l^k} (-1) = -1 \quad (4.76)$$

for all $0 \le j \le M-1$ since not all $q_{i,j}^k = 0$. Hence, by definition, $\lambda_{\max}^k$ is also greater than -1. Since $-1 < \lambda_{\max}^k < 0, 0 < 1 - (\lambda_{\max}^k)^2 < 1$. Under the assumption that the number of iterations is finite, for a fixed number of regions M, the performance of the proposed scheme can be analyzed in the asymptotic regime. Under this assumption, $d_{m,k} = \frac{2N}{M^{k+1}}$ grows linearly with the number of sensors N for $0 \le k \le k^{stop}$. Then,

$$\lim_{N\to\infty} P_D = \lim_{N\to\infty} \prod_{k=0}^{k^{stop}} P_d^k \quad (4.77)$$

$$\ge \prod_{k=0}^{k^{stop}} \lim_{N\to\infty} \left[(1-(M-1)(1-(\lambda_{\max}^k)^2)^{d_{m,k}/2}\right] \quad (4.78)$$

$$= \prod_{k=0}^{k^{stop}} (1-(M-1) \lim_{N\to\infty} \left[(1-(\lambda_{\max}^k)^2)^{d_{m,k}/2}\right] \quad (4.79)$$

$$= \prod_{k=0}^{k^{stop}} [1-(M-1)0] \quad (4.80)$$

$$= \prod_{k=0}^{k^{stop}} 1 = 1. \quad (4.81)$$

Hence, the overall detection probability becomes '1' as the number of sensors N goes to infinity. This shows that the proposed scheme asymptotically attains perfect region detection probability irrespective of the value of finite noise variance. □

Numerical Results

Some numerical results are now presented which justify the analytical results presented so far and provide some insights. It was observed that the performance of the basic scheme quantified by the probability of region detection asymptotically

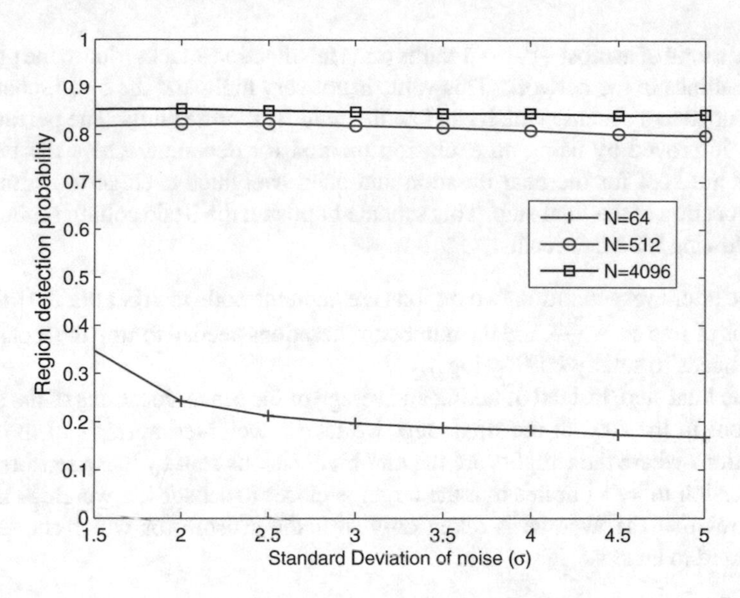

Fig. 4.15 Region detection probability versus the standard deviation of noise with varying number of sensors [20]

approaches '1' irrespective of the finite noise variance. Figure 4.15 shows that the region detection probability increases as the number of sensors approaches infinity. Observe that for a fixed noise variance, the region detection probability increases with an increase in the number of sensors. Also, for a fixed number of sensors, the region detection probability decreases with σ when the number of sensors is small. But when the number of sensors is large, the reduction in region detection probability with σ is negligible and as $N \rightarrow \infty$ the region detection probability converges to 1.

Now consider the case when there are Byzantines in the network. As discussed before, Byzantines are local sensors which send false information to the FC to deteriorate the network's performance. Assume the presence of $B = \alpha N$ number of Byzantines in the network. Here, the Byzantines are assumed to attack the network independently where the Byzantines flip their data with probability '1' before sending it to the FC which has been shown to be the optimal independent attack strategy for the Byzantines. In other words, the data sent by the ith sensor is given by:

$$u_i = \begin{cases} D_i & \text{if } i\text{th sensor is honest} \\ \bar{D}_i & \text{if } i\text{th sensor is Byzantine} \end{cases} \tag{4.82}$$

For such a system, it has been shown that the FC becomes blind to network's information for $\alpha \geq 0.5$. Therefore, the system is analyzed when $\alpha < 0.5$. For the basic scheme described so far which has been designed for an ideal case, each column in C^k contains only one '1' and every row of C^k contains exactly $\frac{N}{M^{k+1}}$ '1's. Therefore, the minimum Hamming distance of C^k is $\frac{2N}{M^{k+1}}$ and, at the $(k+1)$th iteration, it can tolerate a total of at most $\frac{N}{M^{k+1}} - 1$ faults (data falsification attacks) due to the presence of Byzantines in the network. This value is not very high, and the basic scheme can be extended to a scheme which can handle more Byzantine faults. The performance can be improved by using an exclusion method for decoding where the two best regions are kept for the next iteration and using weighted average to estimate the target location at the final step. This scheme builds on the basic coding scheme with the following modifications:

- Since after every iteration two regions are kept, the code matrix after the kth iteration is of size $M \times \frac{2^k N}{M^k}$ and the number of iterations needed to stop the localization task needs to satisfy $k^{stop} < \log_{M/2} N$.
- At the final step, instead of taking an average of the sensor locations of the sensors present in the ROI at the final step, we take a weighted average of the sensor locations where the weights are the one-bit decisions sent by these sensors. Since a decision $u_i = 1$ implies that the target is closer to sensor i, a weighted average ensures that the average is taken only over the sensors for which the target is reported to be close.

Therefore, the target location estimate is given by

$$\hat{\theta}_x = \frac{\sum_{i \in ROI_{k^{stop}}} u_i x_i}{\sum_{i \in ROI_{k^{stop}}} u_i} \tag{4.83}$$

$$\text{and} \quad \hat{\theta}_y = \frac{\sum_{i \in ROI_{k^{stop}}} u_i y_i}{\sum_{i \in ROI_{k^{stop}}} u_i}. \tag{4.84}$$

The exclusion method results in a better performance compared to the basic coding scheme since it keeps the two best regions after every iteration. This observation is also evident in the numerical results presented in [20].

Performance Analysis

Byzantine Fault Tolerance Capability

When the exclusion method-based scheme is used, since the two best regions are considered after every iteration, the fault tolerance performance improves and a total of at most $\frac{2^{k+1} N}{M^{k+1}} - 1$ faults can be tolerated.

Proposition 4.5 ([20]) *The maximum fraction of Byzantines that can be handled at the* $(k+1)$*th iteration by the proposed exclusion method-based coding scheme is limited by* $\alpha_f^k = \frac{2}{M} - \frac{M^k}{2^k N}$.

Proof The proof is straightforward and follows from the fact that the error-correcting capability of the code matrix C^k at $(k+1)$th iteration is at most $\frac{2^{k+1} N}{M^{k+1}} - 1$. Since there are $\frac{2^k N}{M^k}$ sensors present during this iteration, the fraction of Byzantine sensors that can be handled is given by $\alpha_f^k = \frac{2}{M} - \frac{M^k}{2^k N}$. □

The performance bounds on the basic coding scheme can be extended to the exclusion-based coding scheme presented previously. When there are Byzantines in the network, the probabilities $q_{i,j}^k$ of (4.74) become

$$q_{i,j}^k = 1 - \left[(1-\alpha) Q\left(\frac{(\eta_i^k - a_i)}{\sigma}\right) + \alpha \left(1 - Q\left(\frac{(\eta_i^k - a_i)}{\sigma}\right)\right)\right]. \quad (4.85)$$

It was shown in Sect. 4.1.3.1 that the detection probability at every iteration approaches '1' as the number of sensors N goes to infinity. However, this result only holds when the condition in (4.70) is satisfied. Notice that, in the presence of Byzantines,

$$q_{i,j}^k = \begin{cases} (1-\alpha)\left(1 - Q\left(\frac{(\eta_i^k - a_i)}{\sigma}\right)\right) + \alpha Q\left(\frac{(\eta_i^k - a_i)}{\sigma}\right), & \text{for } i \in S_j^k \\ (1-\alpha) Q\left(\frac{(\eta_i^k - a_i)}{\sigma}\right) + \alpha \left(1 - Q\left(\frac{(\eta_i^k - a_i)}{\sigma}\right)\right), & \text{for } i \in S_l^k \end{cases}, \quad (4.86)$$

which can be simplified as

$$q_{i,j}^k = \begin{cases} (1-\alpha) - (1-2\alpha) Q\left(\frac{(\eta_i^k - a_i)}{\sigma}\right), & \text{for } i \in S_j^k \\ \alpha + (1-2\alpha) Q\left(\frac{(\eta_i^k - a_i)}{\sigma}\right), & \text{for } i \in S_l^k \end{cases}. \quad (4.87)$$

Now using the pairwise sum approach presented before, we have:

$$q_{i_j,j}^k + q_{i_l,j}^k = 1 - (1-2\alpha)\left[Q\left(\frac{(\eta - a_{i_j})}{\sigma}\right) - Q\left(\frac{(\eta - a_{i_l})}{\sigma}\right)\right], \quad (4.88)$$

which is an increasing function of α since $Q\left(\frac{(\eta - a_{i_j})}{\sigma}\right) > Q\left(\frac{(\eta - a_{i_l})}{\sigma}\right)$ for all finite σ as discussed before. Therefore, when $\alpha < 0.5$, the pairwise sum in (4.88) is strictly less than 1 and the condition (4.70) is satisfied. However, when $\alpha \geq 0.5$, $\sum_{i \in S_j^k \cup S_l^k} q_{i,j}^k \geq \frac{N_k}{M}$. Therefore, the condition fails when $\alpha \geq 0.5$. It has been shown in Sect. 4.1.2 that the FC becomes blind to the local sensor's information when $\alpha \geq 0.5$. Next, the main theorem for the case when Byzantines are present in the networks is presented.

Theorem 4.3 ([20]) *Let α be the fraction of Byzantines in the networks. Under Assumption 4.1, when $\alpha < 0.5$, $\lim_{N \to \infty} P_D = 1$.*

These derived performance bounds can also be used for system design as shown in [20]. Consider N sensors uniformly deployed in a square region. Let this region be split into M equal regions. From Proposition 4.5, α_f^k is a function of M and N. Also, the detection probability expressions and bounds are functions of M and N. Hence, for given fault tolerance capability and region detection probability requirements, one can find the corresponding number of sensors (N_{req}) to be used and the number of regions to be considered at each iteration (M_{req}). Some guidelines for system design of a network which adopts the proposed approach are presented in the following.

Table 4.2 Target region detection probability and Byzantine fault tolerance capability with varying N ($M = 4$) [20]

N	Target region detection probability	Byzantine fault tolerance capability
32	0.4253	0.4688
128	0.6817	0.4844
512	0.6994	0.4922

Suppose that a system is to be designed that splits into $M = 4$ regions after every iteration. How should a system designer decide the number of sensors N in order to meet the target region detection probability and Byzantine fault tolerance capability requirements? Table 4.2 shows the performance of the system in terms of the target region detection probability and Byzantine fault tolerance capability with varying number of sensors found using the expressions derived in Proposition 4.5.

From Table 4.2, it can be observed that the performance improves with increasing number of sensors. However, as a system designer, one would like to minimize the number of sensors that need to be deployed while assuring a minimum performance guarantee. In this example, if one is interested in achieving a region detection probability of approximately 0.7 and a Byzantine fault tolerance capability close to 0.5, $N = 512$ sensors are sufficient.

Simulation Results

In this section, some simulation results are presented to evaluate the performance of the two schemes in the presence of Byzantine faults. The performance is analyzed using two performance metrics: mean square error (MSE) of the estimated location and probability of detection (P_D) of the target region. A network of $N = 512$ sensors is deployed in a regular 8×8 grid as shown in Fig. 4.13. Let α denote the fraction of Byzantines in the network that is randomly distributed over the network. The received signal amplitude at the local sensors is corrupted by AWGN noise with noise standard deviation $\sigma = 3$. The power at the reference distance is $P_0 = 200$. At every iteration, the ROI is split into $M = 4$ equal regions as shown in Fig. 4.13. We stop the iterations for the basic coding scheme after $k^{stop} = 2$ iterations. The number of sensors in the ROI at the final step is therefore 32. In order to have a fair comparison, we stop the exclusion method after $k^{stop} = 4$ iterations, so that there are again 32 sensors in the ROI at the final step.

Figure 4.16 shows the performance of the two schemes in terms of the MSE of the estimated target location when compared with the traditional maximum likelihood estimation [10]. The MSE has been found by performing 1×10^3 Monte Carlo runs with the true target location randomly chosen in the 8×8 grid.

As can be seen from Fig. 4.16, the performance of the exclusion method-based coding scheme is better than the basic coding scheme and outperforms the traditional MLE-based scheme when $\alpha \leq 0.375$. When $\alpha > 0.375$, the traditional MLE-based scheme has the best performance.

However, it is important to note that the error-correcting code-based schemes provide a coarse estimate as against the traditional MLE-based scheme which optimizes

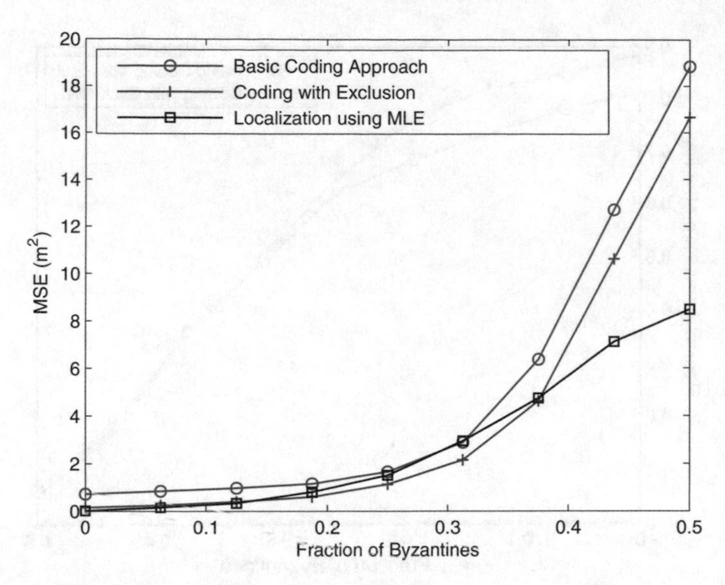

Fig. 4.16 MSE comparison of the three localization schemes: basic coding scheme, coding with exclusion, and ML-based localization [20]

over the entire ROI. Also, the traditional scheme is computationally much more expensive than the coding-based schemes. In the simulations performed in [20], the error-correcting code-based schemes are around 150 times faster than the conventional scheme when the global optimization toolbox in MATLAB was used for the optimization in the ML-based scheme. The computation time is very important in a scenario when the target is moving and a coarse location estimate is needed in a timely manner.

Figure 4.17 shows the performance of the error-correcting code-based schemes in terms of the detection probability of the target region. The detection probability has been found by performing 1×10^4 Monte Carlo runs with the true target randomly chosen in the ROI. Figure 4.17 shows the reduction in the detection probability with an increase in α when more sensors are Byzantines sending false information to the FC.

In order to analyze the effect of the number of sensors on the performance, simulations were performed in [20] by changing the number of sensors and keeping the number of iterations the same as before. According to Proposition 4.5, when $M = 4$, the proposed scheme can asymptotically handle up to 50% of the sensors being Byzantines. Figures 4.18 and 4.19 show the effect of the number of sensors on MSE and detection probability of the target region, respectively, when the exclusion method-based coding scheme is used. As can be seen from Figs. 4.18 and 4.19, the fault tolerance capability of the proposed scheme improves with an increase in the number of sensors and approaches $\alpha_f^k = 0.5$ asymptotically.

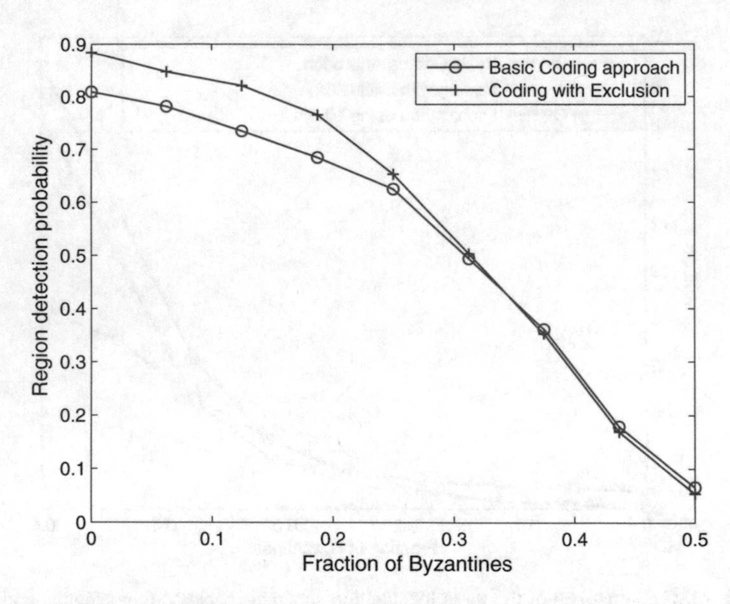

Fig. 4.17 Probability of detection of target region as a function of α [20]

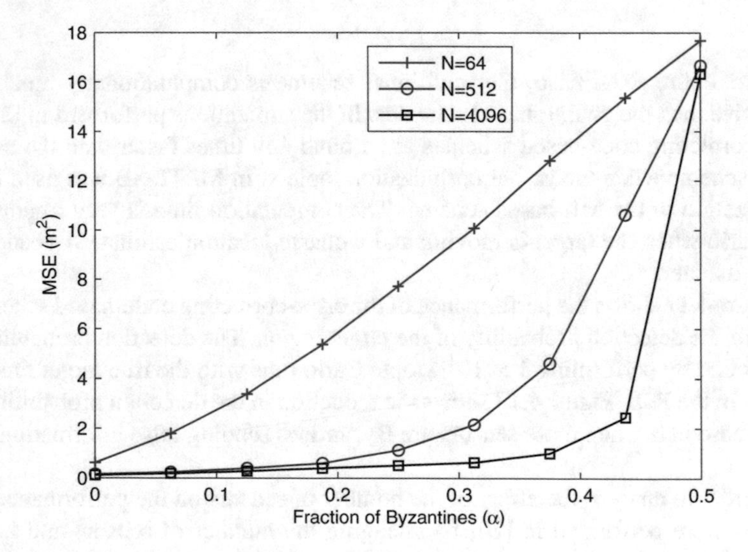

Fig. 4.18 MSE of the target location estimate with varying N [20]

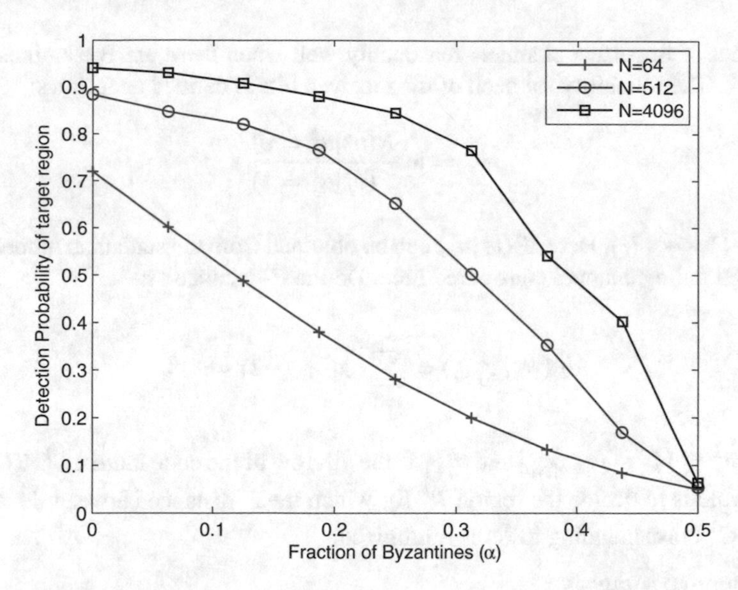

Fig. 4.19 Probability of detection of target region with varying N [20]

Soft-Decision Decoding for Non-ideal Channels

While Byzantines have been considered in the previous section, another practical cause of unreliability in the observations is the presence of imperfect channels. In [20], the scheme was extended to counter the effect of non-ideal channels on system performance. Besides the faults due to the Byzantines in the network, the presence of non-ideal channels further degrades the localization performance. To combat the channel effects, a soft-decision decoding rule was used at every iteration, instead of the minimum Hamming distance decoding rule.

Decoding Rule

At each iteration, the local sensors transmit their local decisions $\boldsymbol{u}^k$ which are possibly corrupted due to the presence of Byzantines. Let the received analog data at the FC be represented as $\boldsymbol{v}^k = [v_1^k, v_2^k, \cdots, v_{N_k}^k]$, where the received observations are related to the transmitted decisions as follows:

$$v_i^k = h_i^k(-1)^{u_i^k}\sqrt{E_b} + n_i^k, \quad \forall i = \{1, \cdots, N_k\}, \tag{4.89}$$

where h_i^k is the fading channel coefficient, E_b is the energy per channel bit, and n_i^k is the additive white Gaussian noise with variance σ_f^2. Here, the channel coefficients are assumed to be Rayleigh distributed with variance σ_h^2.

The FC is assumed to have no knowledge of the fraction of Byzantines α. Hence, a simple reliability measure ψ_i^k was used in the decoding rule in [20] that is not related to local decisions of sensors. It was shown that this reliability measure captures the effect of imperfect channels reasonably well when there are Byzantines in the network. The reliability for each of the received bits is defined as follows:

$$\psi_i^k = \ln \frac{P(v_i^k | u_i^k = 0)}{P(v_i^k | u_i^k = 1)} \tag{4.90}$$

for $i = \{1, \cdots, N\}$. Here, $P(v_i^k | u_i^k)$ can be obtained from the statistical model of the Rayleigh fading channel considered here. Define F-distance as

$$d_F(\boldsymbol{\psi}^k, \boldsymbol{c}_{j+1}^k) = \sum_{i=1}^{N_k} (\psi_i^k - (-1)^{c_{(j+1)i}^k})^2, \tag{4.91}$$

where $\boldsymbol{\psi}^k = [\psi_1^k, \cdots, \psi_{N_k}^k]$ and $\boldsymbol{c}_{j+1}^k$ is the jth row of the code matrix C^k. Then, the fusion rule is to decide the region R_j^k for which the F-distance between $\boldsymbol{\psi}^k$ and the row of C^k corresponding to R_j^k is minimized.

Performance Analysis

Some bounds on the performance of the soft-decision decoding scheme were presented in [20] in terms of the detection probability. Without loss of generality, assume $E_b = 1$. As mentioned before in (4.62), the overall detection probability is the product of the probabilities of detection at each iteration, P_d^k. The following lemma is first presented without proof which was used to prove the theorem stated later.

Lemma 4.3 ([22]) *Let* $\tilde{\psi}_i^k = \psi_i^k - E[\psi_i^k | \theta]$*, and then*

$$E\left[(\tilde{\psi}_i^k)^2 | \theta\right] \le \frac{8}{\sigma^4} \{E[(h_i^k)^4] + E[(h_i^k)^2]\sigma_f^2\}, \tag{4.92}$$

where σ^2 is the variance of the noise at the local sensors whose observations follow (4.2). For the Rayleigh fading channel considered here, both $E[(h_i^k)^4]$ and $E[(h_i^k)^2]$ are bounded and therefore the LHS of (4.92) is also bounded.

Lemma 4.4 ([20]) *Let* $\theta \in R_j^k$ *be the fixed target location. Let* $P_{e,j}^k(\theta)$ *be the error probability of detection of the target region given* $\theta \in R_j^k$ *at the* $(k+1)$*th iteration. For the reliability vector* $\boldsymbol{\psi}^k = [\psi_1^k, \cdots, \psi_{N_k}^k]$ *of the* $N_k = N/M^k$ *observations and code matrix* C^k *used at the* $(k+1)$*th iteration,*

$$P_{e,j}^k(\theta) \le \sum_{0 \le l \le M-1, l \ne j} P\left\{ \sum_{i \in S_j^k \cup S_l^k} Z_i^{jl} \tilde{\psi}_i^k < - \sum_{i \in S_j^k \cup S_l^k} Z_i^{jl} E[\psi_i^k | \theta] \middle| \theta \right\}, \tag{4.93}$$

where $Z_i^{jl} = \frac{1}{2}((-1)^{c_{ji}^k} - (-1)^{c_{li}^k})$.

Proof

$$
\begin{aligned}
P_{e,j}^k(\theta) &= P\{\text{detected region} \neq R_j^k|\theta\} \\
&\leq P\left\{d_F(\boldsymbol{\psi}^k, \mathbf{c}_{j+1}^k) \geq \min_{0\leq l\leq M-1, l\neq j} d_F(\boldsymbol{\psi}^k, \mathbf{c}_{l+1}^k)|\theta\right\} \\
&\leq \sum_{0\leq l\leq M-1, l\neq j} P\left\{d_F(\boldsymbol{\psi}^k, \mathbf{c}_{j+1}^k) \geq d_F(\boldsymbol{\psi}^k, \mathbf{c}_{l+1}^k)|\theta\right\} \\
&= \sum_{0\leq l\leq M-1, l\neq j} P\left\{\sum_{i=1}^{N_k}(\psi_i^k - (-1)^{c_{(j+1)i}^k})^2 \geq (\psi_i^k - (-1)^{c_{(l+1)i}^k})^2|\theta\right\} \\
&= \sum_{0\leq l\leq M-1, l\neq j} P\left\{\sum_{i\in S_j^k\cup S_l^k} Z_i^{jl}\psi_i^k < 0\middle|\theta\right\} \\
&= \sum_{0\leq l\leq M-1, l\neq j} P\left\{\sum_{i\in S_j^k\cup S_l^k} Z_i^{jl}\tilde{\psi}_i^k < -\sum_{i\in S_j^k\cup S_l^k} Z_i^{jl}E[\psi_i^k|\theta]\middle|\theta\right\}.
\end{aligned}
\tag{4.94}
$$

□

Let $\sigma_{\tilde{\psi}}^2(\theta) = \sum_{i\in S_j^k\cup S_l^k} E\left[(Z_i^{jl}\tilde{\psi}_i^k)^2|\theta\right] = \sum_{i\in S_j^k\cup S_l^k} E\left[(\tilde{\psi}_i^k)^2|\theta\right]$, and then the above result can be rewritten as

$$
P_{e,j}^k(\theta) \leq \sum_{0\leq l\leq M-1, l\neq j} P\left\{\frac{1}{\sigma_{\tilde{\psi}}(\theta)}\sum_{i\in S_j^k\cup S_l^k} Z_i^{jl}\tilde{\psi}_i^k < -\frac{1}{\sigma_{\tilde{\psi}}(\theta)}\sum_{i\in S_j^k\cup S_l^k} Z_i^{jl}E[\psi_i^k|\theta]\middle|\theta\right\}.
\tag{4.95}
$$

Under the assumption that for a fixed M, $\frac{N}{M^{k+1}} \to \infty$ as $N \to \infty$ for $k = 0, \cdots, k^{stop}$, we have the following result for asymptotic performance of the proposed soft-decision rule decoding-based scheme.

Theorem 4.4 ([20]) *Under Assumption 4.1, when* $\alpha < 0.5$,

$$
\lim_{N\to\infty} P_D = 1.
$$

Proof The proof is provided in Appendix A.17. □

Note that the detection probability of the proposed scheme can approach '1' even for extremely poor channels with very low channel capacity. This is true because, for fixed M, when N approaches infinity, the code rate of the code matrix approaches zero. Hence, even for extremely poor channels, the code rate is still less than the channel capacity.

Numerical Results

Some numerical results are presented which show the improvement in the system performance when soft-decision decoding rule is used instead of the hard-decision decoding rule in the presence of Byzantines and non-ideal channels. As defined before, α represents the fraction of Byzantines and evaluates the performance of the basic coding approach with soft-decision decoding at the FC. Consider the scenario with the following system parameters: $N = 512$, $M = 4$, $A = 8^2 = 64$ square units, $P_0 = 200$, $\sigma = 3$, $E_b = 1$, $\sigma_f = 3$, and $E[(h_i^k)^2] = 1$ which corresponds to $\sigma_h^2 = 1 - \frac{\pi}{4}$. The basic approach is stopped after $k^{stop} = 2$ iterations. Note that in the presence of non-ideal channels, α_{blind} is less than 0.5 since the non-ideal channels add to the errors at the FC. The number of Byzantine faults which the network can handle reduces and is now less than 0.5. In the simulations, the performance of the schemes deteriorates significantly when $\alpha \to 0.4$ (as opposed to 0.5 observed before), and therefore, the results are only plotted for the case when $\alpha \leq 0.4$.

Figure 4.20 shows the reduction in mean square error when the soft-decision decoding rule is used instead of the hard-decision decoding rule. Similarly, Fig. 4.21 shows the improvement in target region detection probability when using the soft-decision decoding rule. The plots are for 5×10^3 Monte Carlo simulations.

As the figures suggest, the performance deteriorates in the presence of non-ideal channels. Also, the performance worsens with an increase in the number of Byzantines. The performance can be improved by using the exclusion method-based coding approach as discussed before in which two regions are stored after every iteration. Figures 4.22 and 4.23 show this improved performance as compared to the basic

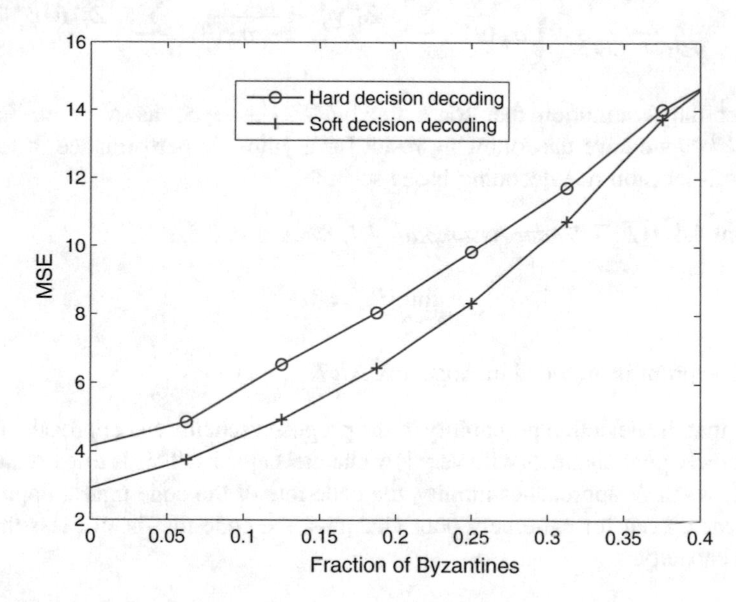

Fig. 4.20 MSE comparison of the basic coding scheme using soft- and hard-decision decoding [20]

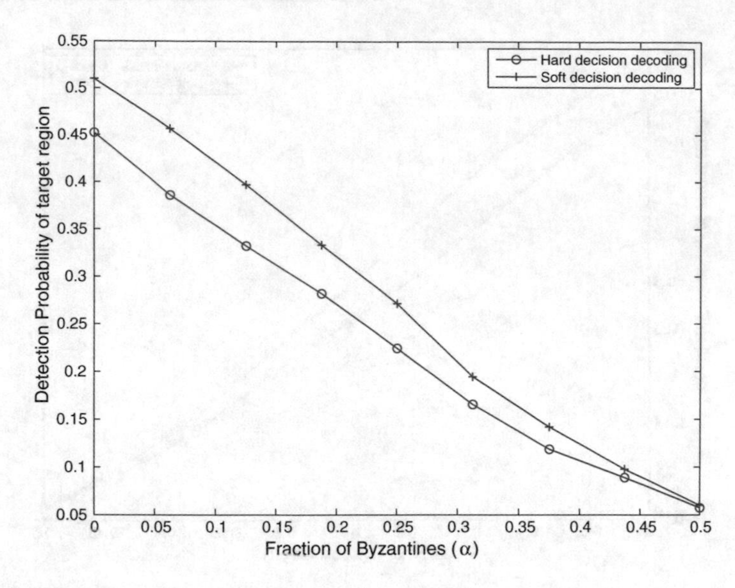

Fig. 4.21 Probability of detection of target region comparison of the basic coding scheme using soft- and hard-decision decoding [20]

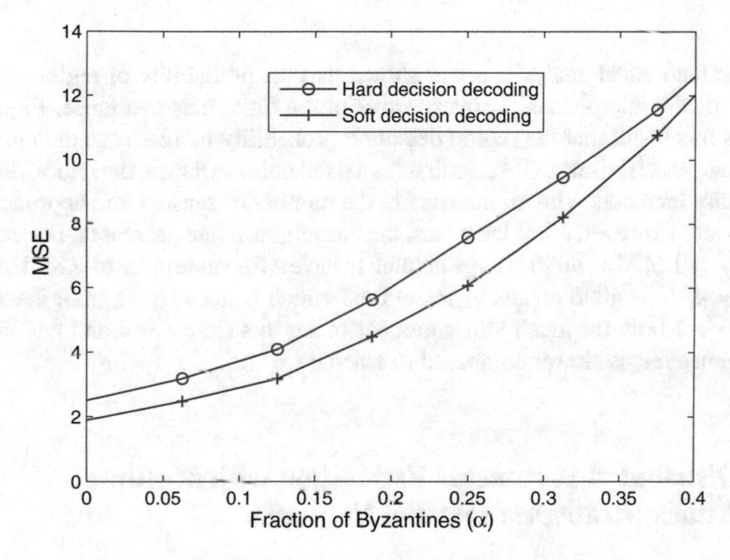

Fig. 4.22 MSE comparison of the exclusion coding scheme using soft- and hard-decision decoding [20]

approach. Note that the exclusion method-based approach also follows the same trend as the basic coding approach with soft-decision decoding performing better than hard-decision decoding.

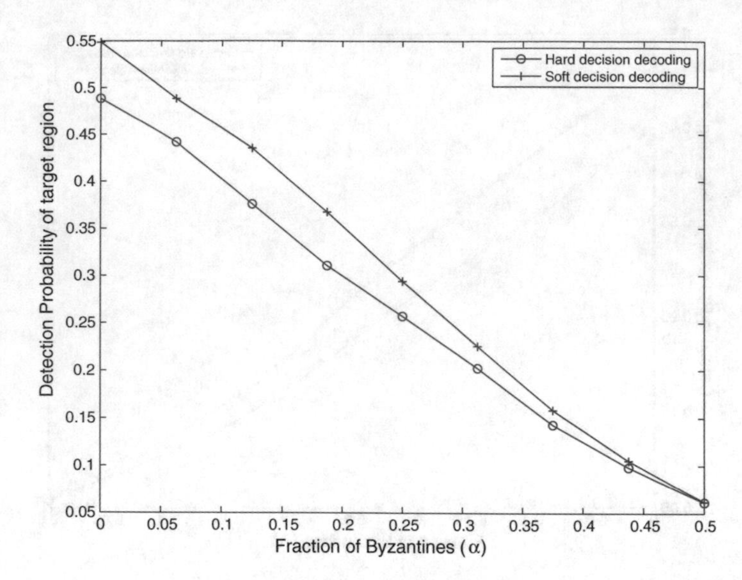

Fig. 4.23 Probability of detection of target region comparison of the exclusion coding scheme using soft- and hard-decision decoding [20]

In the theoretical analysis, it was shown that the probability of region detection asymptotically approaches '1' irrespective of the finite noise variance. Figure 4.24 presents this result that the region detection probability increases as the number of sensors approach infinity. Observe that for a fixed noise variance, the region detection probability increases with an increase in the number of sensors and approaches '1' as $N \rightarrow \infty$. However, as σ_f increases, the convergence rate decreases. For example, when $\sigma_f = 1.5$, $N = 4096$ is large enough to have a P_D value close to 0.85. However, for $\sigma_f = 4$, $N = 4096$ results in $P_D = 0.65$ which is not very large. It is expected that $P_D \rightarrow 1$ only for much larger number of sensors for $\sigma_f = 4$ and therefore the convergence rate is slower compared to when $\sigma_f = 1.5$.

4.2 Distributed Parameter Estimation with Multiple Attack Strategies: Parallel Networks

We now consider the generic parameter estimation problem in a parallel sensor network where there are multiple subsets of attack sensors attacking with different attack parameters. This problem was considered in [24], and it was shown that the attacking sensors can be perfectly categorized as the number of observations per sensor or number of sensors in the network approach infinity. Based on this categorization, a joint estimation scheme was also presented in [24] to estimate the attack parameters

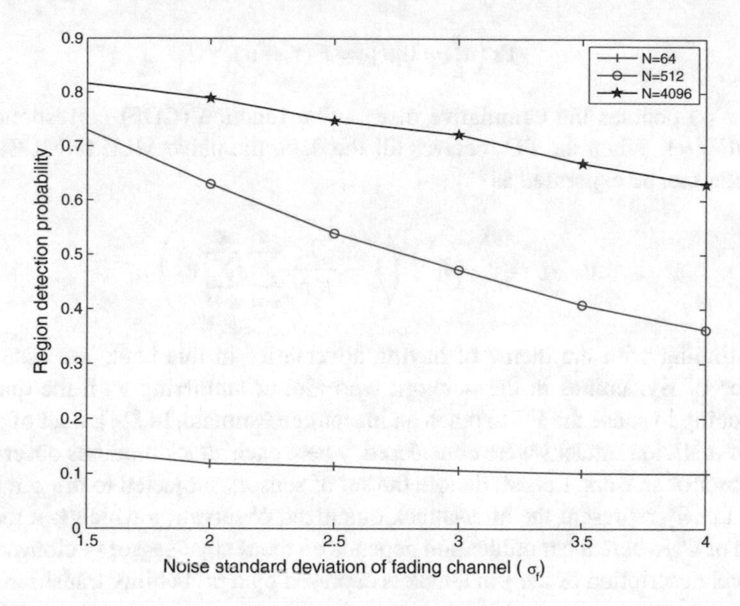

Fig. 4.24 Probability of detection of target region of the exclusive coding scheme using soft-decision decoding with varying number of sensors [20]

along with the unknown parameter. To make this proposed scheme more effective, the quantization scheme was re-designed in [24].

4.2.1 System Model

Consider a set of N distributed sensors, each making observations of a deterministic scalar parameter θ corrupted by additive noise. At the ith sensor, the observation at the kth time instant is described by

$$x_i^k = \theta + n_i^k, \tag{4.96}$$

where n_i^k denotes the additive noise sample with identical zero-mean continuous PDF $f(n_i^k)$ at each sensor and where $\{n_i^k\}$ are i.i.d. It is assumed that the sensors use static and identical thresholds τ for quantization and send binary data to the FC. The quantized data u_j^k is given by

$$u_i^k = \mathbb{I}\{x_i^k \in (\tau, \infty)\}. \tag{4.97}$$

Assume that the quantizer parameter, i.e., the threshold, is known to the FC. The PMF of each u_j^k (under no attack) is identical and is given by

$$\Pr\left(u_i^k = 0|\theta\right) = F(\tau - \theta), \tag{4.98}$$

where $F(\cdot)$ denotes the cumulative distribution function (CDF) corresponding to the PDF $f(\cdot)$. When the FC receives all the data, the naive MLE (NMLE) of the parameter can be expressed as

$$\hat{\theta}_{NML} = \tau - F^{-1}\left(1 - \frac{1}{KN}\sum_{i=1}^{N}\sum_{k=1}^{K} u_i^k\right). \tag{4.99}$$

Continuing with the theme of having adversaries in this book, we assume the presence of Byzantines in the network who aim at tampering with the quantized data, hoping to cause the FC to reach an inaccurate estimate. In [24], a set of distinct types of malicious attacks were considered, where each attack modifies observations of a subset of sensors. Let $\mathscr{A}_p$ denote the set of sensors subjected to the pth type of attack. Let $\tilde{u}_i^k$ represent the after-attack quantized observation which is a modified version of u_i^k, where the modification depends on the set that sensor i belongs to. The statistical description of the pth attack is captured by a probability transition matrix ψ_p,

$$\psi_p = \begin{bmatrix} \psi_{p,0} & 1 - \psi_{p,1} \\ 1 - \psi_{p,0} & \psi_{p,1} \end{bmatrix}, \tag{4.100}$$

which consists of the flipping probabilities $(\psi_{p,0}, \psi_{p,1})$, where $\psi_{p,0} = \Pr\left(\tilde{u}_i^k = 0|u_i^k = 0\right)$ and $\psi_{p,1} = \Pr\left(\tilde{u}_i^k = 1|u_i^k = 1\right)$. Also, the number of attacking sensors in each set $\mathscr{A}_p$ is assumed to be at least Δ ($\Delta > 0$) percent of the total number of sensors.

4.2.2 *Identification and Categorization of Sensors*

Based on the above problem setup, a hypothesis testing problem was formulated in [24]. This test, performed for a given subset of sensors, determines if this set of sensors are *homogenous* or *non-homogeneous*. Homogeneous sensors are those that have the same statistical properties. In other words, all these sensors belong to the same set $\mathscr{A}_p$. On the other hand, the set of non-homogeneous sensors comprises of sensors from different homogeneous sets of sensors. This hypothesis testing problem can be used across all possible subsets of sensors, to identify and categorize sensors into their corresponding sets $\mathscr{A}_p$. However, the efficacy of such a scheme only depends on the performance of the underlying hypothesis testing problem. In [24], the performance of this categorization scheme was characterized in terms of the error probability of the underlying hypothesis testing problem. It was shown that the error probability asymptotically approaches 0 as $N \to \infty$ or $K \to \infty$. In this section, we summarize the major results of [24] and outline their approach.

Let $\mathscr{A}_0$ represent the set of all honest sensors. Therefore, all the sensors in the network can be split into $P+1$ sets $\{\mathscr{A}_p\}$ for $p = 0, \ldots, P$. And let $S_T = \cup_{p=0}^{P} \mathscr{A}_p$ denote the set of all the sensors in the network. Define the following two collections of sets of sensors, whose elements are the subsets of S_T:

$$\mathscr{C}_0 \triangleq \{S \subset S_T | \exists p \text{ s.t.} S \subset \mathscr{A}_p\}, \tag{4.101}$$

$$\mathscr{C}_1(\kappa) \triangleq \{S \subset S_T | \exists S_1, S_2 \subset S \text{ and } S_1, S_2 \subset \mathscr{C}_0, \text{s.t.} P_{S_1} \geq \Delta,\ P_{S_2} \geq \Delta, \text{ and } S_1 \cup S_2 \notin \mathscr{C}_0\}, \tag{4.102}$$

where $\kappa \in (0, \Delta)$ and P_S represents the percentage of sensors in S relative to the sensor network.

From these definitions, $\mathscr{C}_0$ denotes the collection of all homogeneous sets of sensors which are statistically identical and $\mathscr{C}_1(\kappa)$ denotes the collection of all non-homogeneous sets of sensors which each occupy more than a given percentage κ of all sensors. Clearly, one can observe that every subset of $\mathscr{A}_p$ for any p is an example of elements in $\mathscr{C}_0$, and a combination of sensors from two distinct sets $\mathscr{A}_p$ that include more than κ percent of the sensors in the network is an example of elements in $\mathscr{C}_1(\kappa)$.

Now for a given subset of sensors $\mathscr{J}$ where $P_{\mathscr{J}} \geq \Delta$, consider the following hypothesis testing problem,

$$\begin{aligned} H_0 &: \mathscr{J} \in \mathscr{C}_0 \\ H_1 &: \mathscr{J} \in \mathscr{C}_1(\kappa). \end{aligned} \tag{4.103}$$

Under certain practical assumptions on the attack strategies and when K is greater than a constant K^*, it has been shown in [24] that the error probability of the hypothesis testing problem in (4.103) is bounded as follows

$$P_{error} \triangleq \pi_0 \Pr(\text{Declare } H_1 | H_0) + \pi_1 \Pr(\text{Declare } H_0 | H_1) \leq C e^{\kappa \gamma K N}, \tag{4.104}$$

where π_0, π_1 are the prior probabilities of the hypothesis testing problem of (4.103), γ is a negative constant, and C is a positive constant.

From this result, we can observe that the error probability in (4.104) decreases to 0 as either $K \to \infty$ or $N \to \infty$, which implies that both enlarging the size of the sensor network and/or increasing the number of time observations at each sensor can improve the FC's ability to determine whether a given set of sensors is homogeneous or not. Based on this fact, further analysis on identification and categorization was conducted in [24] for each of the cases when $K \to \infty$ or $N \to \infty$, and the major results are stated here without proofs.

K is Sufficiently Large

Theorem 4.5 ([24]) *For any N as $K \to \infty$, the FC can always identify from the observations, without further knowledge, a group of sensors which make up $P_{\mathscr{A}_0}$ percent of all the sensors such that this group contains zero percent attacked sensors. In this sense, one can always identify the unattacked sensors. Moreover, as $K \to \infty$, the FC is also able to identify the other P groups of sensors, which, respectively,* *make up* $\{P_{\mathscr{A}_p}\}_{p=1}^{P}$ *percent of all the sensors, such that for* $p = 1, \ldots, P$*, group* p *contains zero percent sensors not experiencing attack* p*.*

From this theorem, the FC can perfectly identify the set of unattacked sensors and categorize the attacked sensors into different groups according to their distinct types of attacks as the number of time samples at each sensor approaches infinity.

N is Sufficiently Large, While K is Finite

Theorem 4.6 ([24]) *Consider a sensor network with* N *sensors, and each sensor observes a finite number* K *of time samples which satisfies* $K \geq K^*$*. As* $N \to \infty$*, the FC can determine the number of attacks in the sensor network. Moreover, for the* p*th attack* $\forall p \geq 0$*, the FC can identify a corresponding group of sensors* $\tilde{\mathscr{A}}_p$ *which closely approximates* $\mathscr{A}_p$*.*

This theorem shows that with a finite number of time samples at each sensor K, as the number of sensors N in the sensor network grows to infinity, the FC is able to ascertain the number of attacks in the sensor network. However, there is no guarantee that the FC can perfectly categorize the sensors into different groups according to distinct attack types.

Also, Theorem 2 in [24] provides an upper bound which quantifies the maximum percentage of sensors that are misclassified in finding every $\mathscr{A}_p$. As stipulated in [24], this upper bound depends on K and monotonically decreases to 0 as K increases to infinity.

Numerical Results

We now present some examples that test the performance of the identification and categorization technique. Consider a sensor network consisting of $N = 10$ sensors, attacked by $P = 2$ attack strategies. The two attacks control 30% and 20% of sensors, respectively, i.e., $P_{\mathscr{A}_1} = 0.3$ and $P_{\mathscr{A}_2} = 0.2$. Let the attack parameters be $(\psi_{1,0}, \psi_{1,1}) = (0.2, 0.8)$ and $(\psi_{2,0}, \psi_{2,1}) = (0.7, 0.1)$, and the parameter to be estimated $\theta = 1$. Also, let $\tau = 1$, $\Delta = 20\%$, and the additive noise obeys a standard normal distribution. Figure 4.25 depicts the percentage of all miscategorized sensors (across $N_{mc} = 200$ Monte Carlo approximations) as a function of the number of time samples at each sensor, K. As observed from the theoretical analysis, there is a diminishing trend of the average percentage of miscategorized sensors and approaches 0 as K increases. This corroborates the theoretical results that the FC can identify and categorize the attacked sensors into different groups according to distinct types of attacks to achieve any desired level of accuracy for a sufficiently large K.

4.2.3 Joint Estimation Including Attack Parameters

In the previous section, we discussed how sensors can be categorized/identified as attacking or non-attacking. While the attacked sensors can be removed from the inference process after identification, a smarter solution was discussed in [24], where

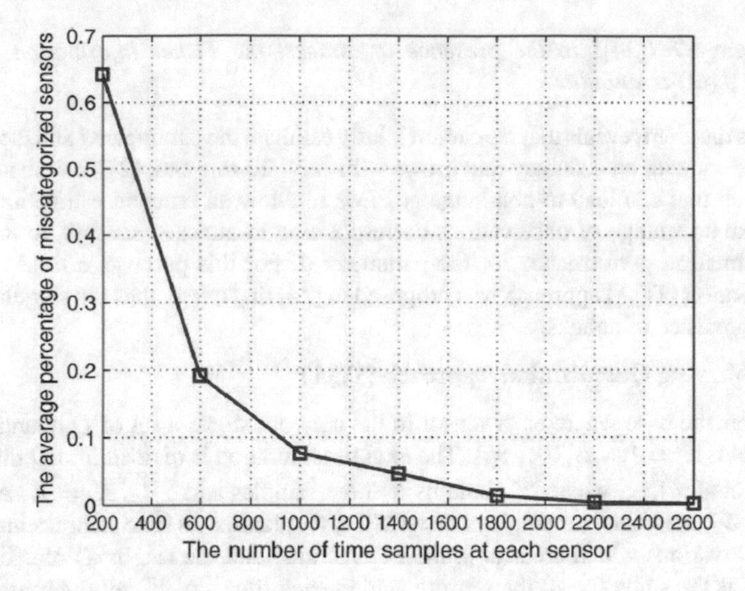

Fig. 4.25 Identification and categorization of attacked sensors [24]

the parameters of the attacked sensors can be learnt and leveraged for further inference using these sensors. Let Θ denote a vector containing the parameter θ along with all the unknown parameters of the attacks

$$\Theta \triangleq [\theta, \psi_{1,0}, \psi_{1,1}, \cdots, \psi_{P,0}, \psi_{P,1}]^T. \tag{4.105}$$

The log-likelihood function evaluated at $\tilde{\mathbf{u}} = \mathbf{r}$ is

$$L(\Theta) = \log \Pr(\tilde{\mathbf{u}} = \mathbf{r}|\Theta) \tag{4.106}$$

$$= \sum_{p=0}^{P} \sum_{i \in \mathscr{A}_p} \sum_{k=1}^{K} \sum_{r_i'^k=0}^{1} \mathbb{I}\{r_i^k = r_i'^k\} \log \Pr\left(\tilde{u}_i^k = r_i'^k|\Theta\right). \tag{4.107}$$

Let $\tilde{p}(\psi_p, \theta) \triangleq \Pr\left(\tilde{u}_i^k = 1|\Theta\right)$, $\forall i \in \mathscr{A}_p$. Then, the Fisher information can be derived as in [24] as follows:

$$J(\Theta) = KN \sum_{p=0}^{P} \zeta_p \Phi_p, \tag{4.108}$$

where $\zeta_p \triangleq \frac{P_{\mathscr{A}p}}{\tilde{p}(\psi_p,\theta)[1-\tilde{p}(\psi_p,\theta)]}$ and $\Phi_p \triangleq \frac{\partial \tilde{p}(\psi_p,\theta)}{\partial \Theta}\left(\frac{\partial \tilde{p}(\psi_p,\theta)}{\partial \Theta}\right)^T$.

Theorem 4.7 ([24]) *In the presence of attacks, the Fisher information matrix (FIM) $J(\Theta)$ is singular.*

This theorem reveals that we cannot jointly estimate the parameter θ and the attack parameters with an accuracy that grows with KN. To this end, [24] investigated an approach that can lead to non-singular FIMs to allow an efficient estimation of Θ and take advantage of observations corresponding to attacked sensors, to improve the estimation performance for the parameter θ. For this purpose, a time-varying quantization (TQA) approach was proposed in [24] that overcomes the singular FIM in the presence of attacks.

Time-Varying Quantization Approach (TQA)

Consider the case where each sensor in the network stores a set of Q quantization thresholds $\mathcal{Q} = \{\tau_1, \tau_2, \ldots, \tau_Q\}$. The total time duration is divided into Q different time slots $\{\mathcal{T}_t\}_{t=1}^{Q}$, where $\mathcal{T}_t$ contains K_t time samples and $\sum_{t=1}^{T} K_t = K$. At time instant $\mathcal{T}_t$, the quantizer employs threshold τ_t to quantize its time samples into one-bit observations which are sent to the FC. Assume that the length K_t of each time slot $\mathcal{T}_t$ is the same for all the sensors, and in each time slot $\mathcal{T}_t$, all the sensors use the same threshold τ_t to quantize their time samples.

Based on the above approach, the following result was derived in [24].

Theorem 4.8 ([24]) *If $Q \geq 2$, then the FIM $J_{TQA}(\Theta)$ for Θ in the TQA is non-singular for any value of θ.*

This result implies that as long as $K \geq 2$, the time samples can be divided into at least two time slots, employ distinct thresholds to quantize the time samples in different time slots, and hence achieve non-singular FIM. Therefore, it is possible for the FC to jointly estimate the parameter θ and the attack parameters simultaneously when the quantized observations are generated by at least two distinct thresholds.

It has also been shown in [24] that the asymptotic performance, characterized by the Cramér–Rao bound (CRB), can be improved by utilizing the observations from the set of attacked sensors in the TQA approach against the simple estimation approach (SEA), under certain conditions. In the following, this performance is numerically characterized via examples.

Consider the same network described before where there are $N = 10$ sensors. However, to simplify the situation and observe the relative CRB gain as a function of attacked sensors, consider a scenario where $P = 1$, and the parameter to be estimated is $\theta = 2$. Each sensor observes $Q = 801$ time slots of times samples which are quantized by employing the threshold set $\mathcal{Q}$. Figure 4.26 presents the relative CRB gain as the percentage of the compromised sensors varies from 0 to 45% for different statistical attack matrices. It can be observed that the relative CRB gain increases with the percentage of attacked sensors for all four cases considered.

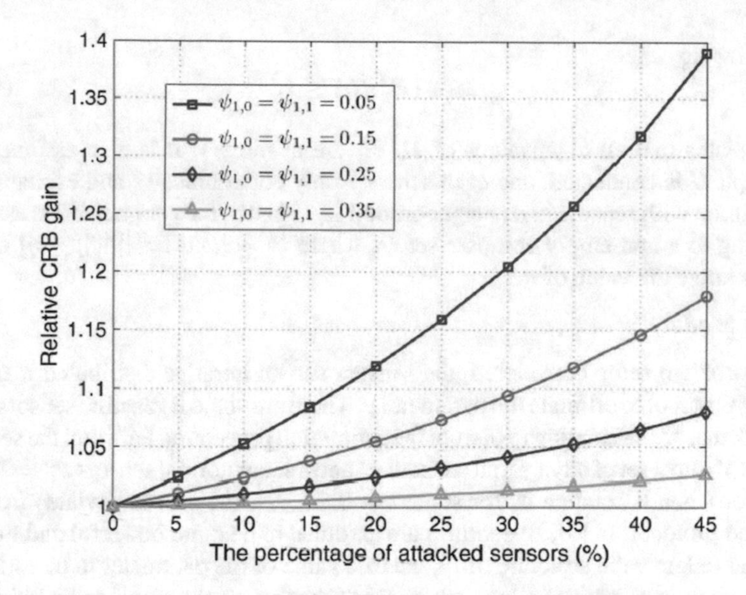

Fig. 4.26 Relative CRB gain versus the percentage of attacked sensors. [24]

4.3 Distributed Parameter Estimation with Adversaries: Peer-to-Peer Networks

In [3], the distributed parameter estimation problem was studied in a peer-to-peer network setup when a subset of the sensors is adversarial. They presented a Flag Raising Distributed Estimator (FRDE) algorithm that handles the attacks and ensures accurate parameter estimation while also detecting the adversarial agents. This algorithm is a consensus+innovations estimator [6] in which agents combine estimates of neighboring agents (consensus) along with their local sensing information (innovations).

4.3.1 System Model

Consider a network of N sensors which communicate without an FC, and their communication is characterized by a communication graph $G = (V, E)$, where V is the set of sensors and E is the set of edges. Let $\theta \in \mathbb{R}^M$ be a deterministic unknown parameter that the N sensors are trying to estimate. Sensor i makes its own measurement at time t as

$$y_i(t) = H_i\theta + w_i(t), \tag{4.109}$$

where H_i is the measurement matrix at the ith sensor and $w_i(t)$ is the zero-mean noise which is assumed to be i.i.d. across time and independent across sensors. Let $E\left[w_i(t)w_i(t)^T\right] = \Sigma_i$ be the covariance matrix. The measurement matrix H_i satisfies the following

$$\lambda_{max}\left(H_i^T H_i\right) \leq 1, \tag{4.110}$$

which states that all eigenvalues of $H_i^T H_i$ are at most 1. It is also assumed that the graph G is connected, and each sensor i only communicates and exchanges its information with sensors in its neighborhood Ω_i. Finally, the parameter θ is assumed to belong to a non-empty compact set Θ, where $\Theta = \{\omega \in \mathbb{R}^M |||w|| \leq \eta\}$ and all sensors know the value of η.

Attack Model

In the problem setup considered, the sensors use an iterative distributed message-passing protocol to estimate the parameter θ. There are some Byzantine sensors in the network that are attempting to disrupt the estimation procedure. Partition the set of all sensors V into a set of adversarial sensors $\mathscr{A}$ and a set of normal sensors $\mathscr{N} = V \backslash \mathscr{A}$. A sensor i is a Byzantine if, for some $t = 0, 1, \ldots,$, the sensor deviates from the designed protocol. In [3], Byzantines are assumed to be quite powerful and they are assumed to know the structure of G, the true value of the parameter to be estimated θ, the true membership of all the sensors (in other words the membership of $\mathscr{A}$ and $\mathscr{N}$). It is also assumed that these Byzantines can communicate among themselves to launch powerful attacks and send arbitrary messages to their neighbors. In [3], for analysis purposes, it was assumed that the induced subgraph $G_{\mathscr{N}}$ of the normally behaving agents is connected.

4.3.2 *Analysis from Network Designer's Perspective*

In this section, we describe the Flag Raising Distributed Estimator (FRDE) algorithm developed in [3] that enables simultaneous parameter estimation and adversary detection, in the above described powerful attack model.

The FRDE algorithm proposed in [3] is iterative, and in each iteration, there are three main steps:

1. Message passing,
2. Estimate update, and
3. Adversary detection.

At each $t = 0, 1, \ldots$, each agent i computes a local estimate $x_i(t)$ of the unknown parameter θ along with a flag value $\pi_i(t)$. This binary flag represents the ith sensor's belief on whether the network is under attack. The distributed algorithm is said to have detected a Byzantine in the network if for some t and i, $\pi_i(t)$ =attack, and we say that the distributed algorithm has not detected any Byzantine if, for all t and all i, $\pi_i(t)$ = no attack. Note that the flag $\pi_i(t)$ = 'attack' only indicates that there exists a Byzantine in the network but the exact identity of the Byzantines is unknown.

The algorithm is initialized at $t = 0$ by setting the estimates of all normal sensors as $x_i(0) = 0$ and the flags at all normal sensors as $\pi_i(0)$ =no attack. Note that the adversarial sensors need not initialize their estimates as prescribed by the algorithm.

At every iteration, the following steps are undertaken:

Message Passing

At every $t = 0, 1, 2, \ldots$, each normal sensor $i \in \mathscr{N}$ generates messages as

$$m_{i,l}^t = x_i(t), \tag{4.111}$$

which are sent to each of its neighbors, $l \in \Omega_i$.

Estimate Update

Based on the messages received, each sensor updates its estimate along with its own observation. For this purpose, each sensor maintains a running average of its local measurements $y_i(t)$ as

$$\bar{y}_i(t) = \frac{t}{t+1}\bar{y}_i(t-1) + \frac{1}{t+1}y_i(t), \tag{4.112}$$

and $\bar{y}_i(0) = y_i(0)$.

Such an averaging ensures that

$$\bar{y}_i(t) = H_i\theta + \frac{1}{t+1}\sum_{j=0}^{t} w_i(j). \tag{4.113}$$

Each honest/normal sensor $i \in \mathscr{N}$ uses the consensus+innovations estimate update rule to update its local estimate using messages received from its neighbors and its own running average as

$$x_i(t+1) = x_i(t) - \beta \sum_{l \in \Omega_i} \left(x_i(t) - m_{l,i}^t\right) + \alpha H_i^T \left(\bar{y}_i(t) - H_i x_i(t)\right), \tag{4.114}$$

where α and β are positive constants that are parameters of the algorithm.

Adversary Detection

At every iteration t, every normal/honest sensor $i \in \mathscr{N}$ updates its flag value as follows:

$$\pi_i(t+1) = \begin{cases} \text{Attack}, & \pi_i(t) = \text{Attack, or } \exists l,\ ||x_i(t) - m_{l,i}^t|| > \gamma_t \\ \text{No Attack}, & \text{Otherwise,} \end{cases} \tag{4.115}$$

where γ_t is a time-varying parameter updated as

$$\gamma_{t+1} = (1-\delta)\gamma_t + \alpha \frac{2K}{(t+1)^{\tau}}, \tag{4.116}$$

and $\gamma_0 = 2\eta\sqrt{N}$. Here, $K > 0$, $0 < \tau < \frac{1}{2}$, and $0 < \delta \leq 1$ are parameters of the algorithm. These parameters $(\alpha,\beta,K,\tau,\delta)$ are selected so as to satisfy some graph-theoretical constraints as described in [3]. A few approaches to select these parameters are also presented in [3]. It is important to note that it is assumed that no Byzantine intentionally reports an attack, i.e., for all $i \in \mathscr{A}$, and for all t, $\pi_i(t) =$ no attack, since they want to avoid being detected.

The performance of the FRDE algorithm in terms of its estimation accuracy and adversary detection capability is analyzed in detail in [3]. Here, we summarize the major results and point the reader to [3] for detailed proofs.

The FRDE algorithm is first analyzed in the ideal setup when there are no Byzantines in the network. For this setup, an upper bound on the false alarm probability that FRDE declares that there is a Byzantine when there are none is derived in [3] along with an analysis on the behavior of local estimates $x_i(t)$ when there are no adversarial agents.

Theorem 4.9 ([3]) *If there are no Byzantines in the network ($\mathscr{A} = \emptyset$), then, under the FRDE algorithm, for all $i \in V$,*

$$P\left(\lim_{t\to\infty}(t+1)^{\tau_0}||x_i(t)-\theta|| = 0\right) = 1, \tag{4.117}$$

for every $0 \leq \tau_0 < \frac{1}{2}$. Moreover, the false alarm probability, P_{FA}, satisfies

$$P_{FA} = P\left(\exists i \in V, t \geq 0 : \pi_i(t) = Attack\right) \leq \frac{\Psi\zeta(\tau)}{K^2}, \tag{4.118}$$

where $\Psi = \sum_{i=1}^{N} trace(\Sigma_i)$ and $\zeta(\tau) = \sum_{j=1}^{\infty} \frac{1}{j^{2(1-\tau)}}$.

This theorem from [3] states that, when there are no Byzantines, local estimates are strongly consistent and almost surely converge to the true parameter value θ. It also bounds the probability that any sensor i at any time $t \geq 0$ raises an attack flag. Therefore, this bounds the overall false alarm probability of the FRDE algorithm.

Moving to the general case when there are Byzantines present in the network, the performance of the FRDE algorithm clearly depends on the relative strength of the Byzantines and honest sensors. Specifically, [3] shows that the algorithm's performance depends on the notion of observability, which can be defined as a function of the measurement matrices. A subgraph $G_{\mathscr{X}}$ induced by a subset $\mathscr{X} \subset V$ of sensors is globally observable if the matrix $\sum_{i\in\mathscr{X}} H_i^T H_i$ is invertible. Based on this definition, it was shown that if the network of normally behaving sensors is not globally observable, then the Byzantines may disrupt the estimation process while evading detection. Byzantines are said to have evaded detection by the FRDE algorithm if the probability of detecting $\mathscr{A}$ is no greater than the false alarm probability of FRDE (Proposition 1 from [3]).

On the other hand, when the network of normally behaving sensors is globally observable, there are two possible cases: Either at least one honest sensor raises an attack flag (i.e., $\exists t\ 0, i \in \mathcal{N} : \pi_i(t) =$ attack), or no honest sensor ever raises an alarm flag (i.e., for all $i \in \mathcal{N}$ and for all $t \geq 0$, $\pi_i(t) =$ no attack). In the first case, obviously FRDE is successful in identifying the presence of an attack. In the second case, the following theorem from [3] states that the local estimates of normally behaving sensors are still consistent.

Theorem 4.10 ([3]) *Let the set of normally behaving sensors $\mathcal{N}$ be connected, and let $G_{\mathcal{N}}$ be globally observable. If, for all $i \in \mathcal{N}$ and for all $t = 0, 1, \ldots, \pi_i(t) =$ no attack, then, under FRDE, for all $i \in \mathcal{N}$,*

$$P\left(\lim_{t\to\infty} (t+1)^{\tau_0}||x_i(t) - \theta|| = 0\right) = 1, \tag{4.119}$$

for every $0 \leq \tau_0 < \tau$.

This theorem states that when the honest sensors are connected and their models are globally observable and if the Byzantines remain undetected, all normally behaving sensors' local estimates asymptotically converge almost surely to θ.

The performance of the FRDE algorithm was demonstrated in [3] using numerical examples. Consider the network in Fig. 4.27 where there are $N = 500$ sensors deployed in a random network to estimate a three-dimensional parameter θ, which corresponds to the x, y, and z coordinates of a target's location. The unknown parameter θ was chosen at random from a sphere of radius $\eta = 500$ m, and the sensors were placed randomly over a two-dimensional square (100 m by 100 m). Two sensors are assumed to be neighbors (an edge exists between them) if the Euclidean distance between them is less than 10 m.

A subset of 160 sensors was randomly selected to be measuring the z component of the target location (these are represented as red diamonds in Fig. 4.27). These agents have a measurement matrix given by

$$H_{Diamond} = \begin{bmatrix} 0 & 0 & 1 \end{bmatrix}. \tag{4.120}$$

The other sensors, represented by the black dots in Fig. 4.27, measure the x and y components of the target location. These sensors have a measurement matrix given by

$$H_{Circle} = \begin{bmatrix} 1 & 0 & 0 \\ 0 & 1 & 0 \end{bmatrix}. \tag{4.121}$$

The local measurements are assumed to be corrupted by additive noise $w_i(t)$, which is an i.i.d. sequence of Gaussian random variables with zero mean and covariance $\sigma^2 I_{p_l}$, where p_i is the dimension of the measurement $y_i(t)$. For the numerical examples, the covariance value is $\sigma^2 = 60$ and the local signal-to-noise ratio (SNR) is 11 dB.

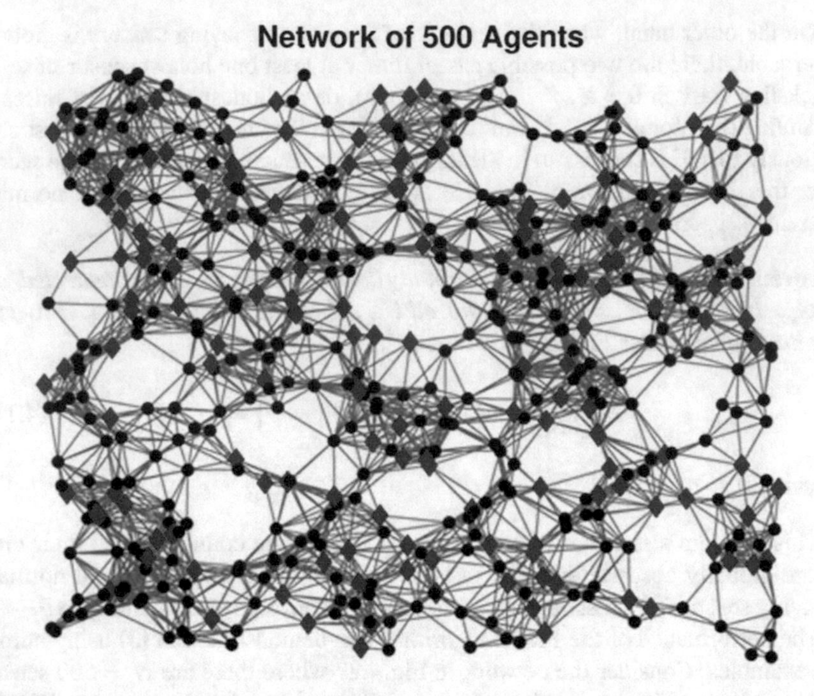

Fig. 4.27 Communication network of the 500 agents. Agents marked by black dots measure the x and y coordinates of the target. Agents marked by red diamonds measure the z coordinate of the target [3]

For the FRDE algorithm, assign the value 0 to the no attack flag and the value 1 to the attack flag. Also, by following the procedure presented in [3], FRDE parameters were picked as $\alpha = 3.1 \times 10^{-2}$, $\beta = 3.1 \times 10^{-2}$, $\delta = 9.0 \times 10^{-3}$, $K = 4$, and $\tau = 0.40$.

To consider the effect of different types of attacks, four different attack strategies were considered in [3]:

1. No adversarial agents: All agents behave normally.
2. Strong adversarial agents: All red diamond sensors are Byzantines and make the remaining normally behaving sensor models globally unobservable.
3. Disruptive weak adversarial agents: Half of the red diamond sensors are Byzantines, perform a disruptive attack, and try to compromise the consistency of the remaining sensors' estimates. In this case, the remaining normally behaving sensor models are globally observable.
4. Undisruptive weak adversarial agents: Half of the red diamond sensors are Byzantines and perform a stealthy attack so that no sensor raises an attack flag. In this case as well, the remaining normally behaving sensor models are globally observable.

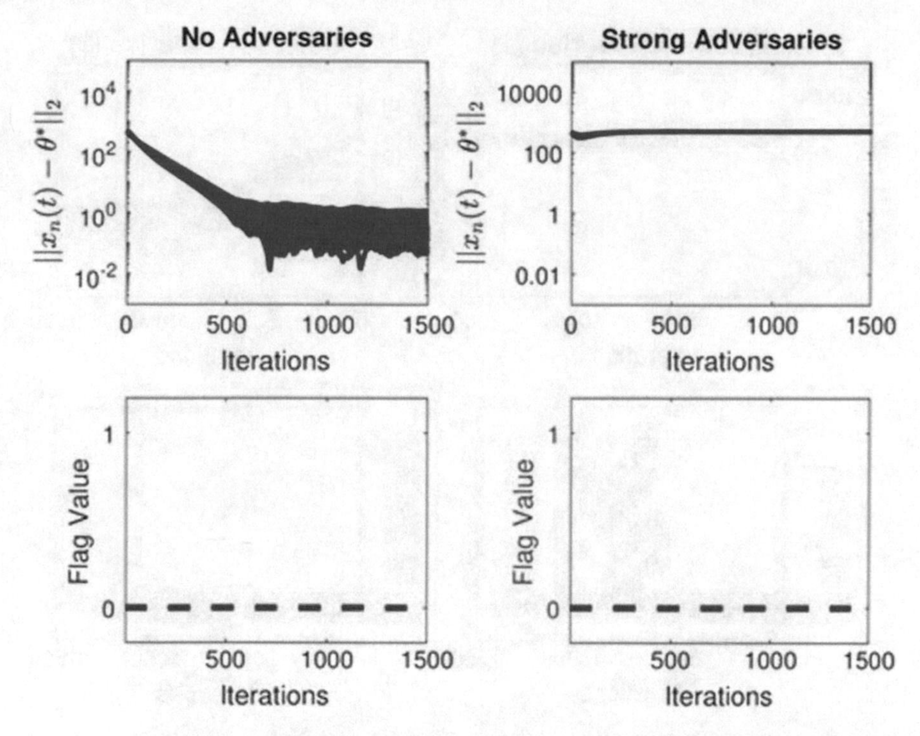

Fig. 4.28 Performance of FRDE when there are no Byzantines (left) and when all 160 red diamond sensors are Byzantines (right). Top: Sensor estimation errors. Bottom: Sensor flag values [3]

As we can observe, the first two models represent the extreme cases of no attack (Theorem 4.9) and full attack causing network of normally behaving sensors to be globally unobservable and causing estimate to be imperfect. On the other hand, the third and fourth attack models are intermediate in nature and cause the network of normally behaving sensors to be globally observable. In this situation too, there are two possibilities: maximum disruption attack (case 3) and stealth attack (case 4).

Figure 4.28 describes the performance of FRDE when all the sensors behave normally and when all red diamond sensors are Byzantines. When all 160 red diamond sensors are adversarial, the normally behaving sensors induce a connected subgraph but are not globally observable. Then, as presented in Proposition 1 in [3], Byzantines can simultaneously avoid detection and cause the normally behaving sensors to estimate θ incorrectly. Figure 4.28 shows Byzantines preventing the network of sensors from converging to the correct estimate while remaining undetected.

For the third and fourth scenarios, half of the red diamonds were randomly selected to be Byzantines. Figure 4.29 shows the effect of an undetectable attack when the normally behaving sensors are globally observable. When the Byzantines try to make the normally behaving sensors converge to an incorrect θ, they are detected by the network (left plots in Fig. 4.29). On the other hand, Byzantines can avoid detection

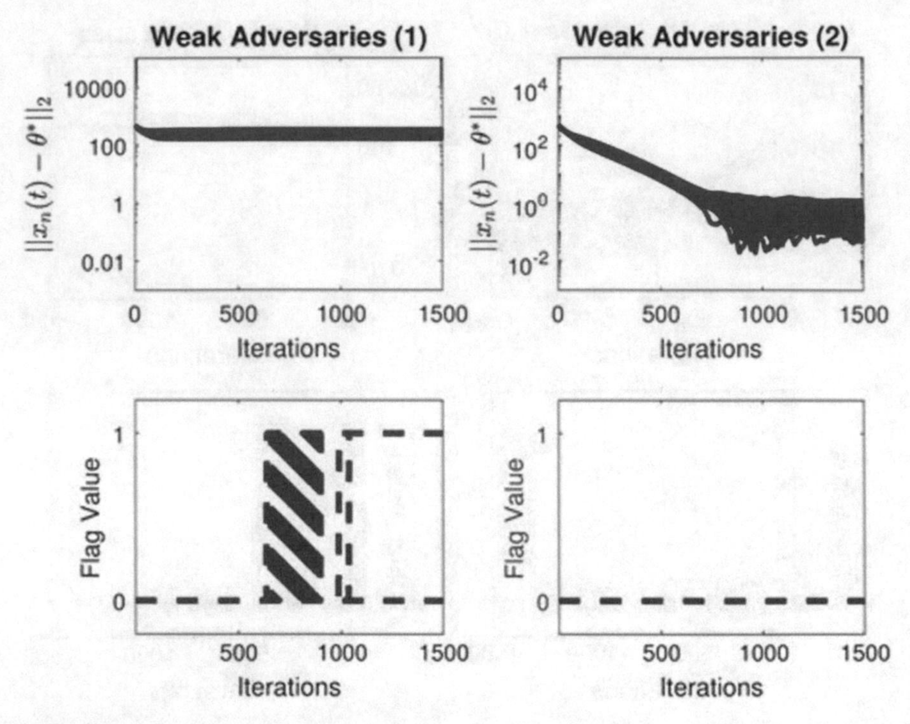

Fig. 4.29 Performance of FRDE when 80 out of 160 red diamond agents are Byzantines. Left: disruptive attack. Right: undisruptive attack. Top: sensor estimation e. Bottom: sensor flag values [3]

in the right plots of Fig. 4.29, as no sensor raises a flag indicating the presence of an adversarial sensor, but the Byzantines cannot prevent the normally behaving sensors from converging to the correct estimate of θ. The results of the third and fourth numerical examples corroborate Theorem 4.10. If the network of normally behaving sensors is connected and globally observable, and if the Byzantines behave in an undetectable manner, the normally behaving sensors' estimates converge to the true parameter θ.

4.4 Summary

In this chapter, we discussed the distributed estimation problem in the presence of Byzantines. First, the problem of reliable localization in sensor networks was explored when the network consists of some Byzantines. We presented the problem from Byzantine's perspective and derived the optimal attack strategies for the attacker who intends to deteriorate the estimation performance at the FC. It was shown that when the fraction of Byzantines in the network is greater than 0.5 and is attacking independently, the FC becomes blind to the data from the sensors and can only estimate the location of the target using the prior information. We then considered the problem from the network designer's perspective and discussed three major schemes proposed in the literature. Each of these schemes is of increasing complexity and proposes more changes in the system mechanisms. The first scheme deals with the identification of the Byzantines by observing the data over time. It was shown that the proposed scheme works well and identifies most of the Byzantines. In order to improve the performance further, the second scheme moves from the traditional static identical thresholds at the local sensors to dynamic nonidentical thresholds. By doing so, one can improve the system performance while also making the Byzantines ineffective in their attack strategy. Both these schemes deal with maximum likelihood type optimal estimators at the FC. The third scheme proposed a sub-optimal but simple and effective estimator at the FC that is based on error-correcting codes. This scheme is shown to be computationally more efficient while also making it robust to Byzantine attacks due to the use of error-correcting codes.

In the second part, results from [24] were presented that studied the distributed estimation problem using quantized data in the presence of multiple Byzantine attacks. An algorithm was presented that can identify the attacked sensors and categorize these sensors into different subsets according to distinct types of attacks *almost* perfectly. Based on the identified sensors, joint estimation of the attack parameters and the unknown parameter to be estimated was considered. However, on observing that the corresponding FIM is singular, a time-varying quantization approach TQA was proposed which divides the observation time interval at each sensor into several time slots and employs distinct thresholds to quantize the time samples in different time slots. It was shown that TQA ensures a non-singular FIM and also improves the CRB for the parameter to be estimated.

While most of the results in distributed estimation with Byzantines have focused on parallel networks, in the final part, we presented results from [3] that studied resilient distributed estimation in a peer-to-peer network. An algorithm was presented, termed Flag Raising Distributed Estimation (FRDE), that allows a network of sensors to reliably estimate an unknown parameter in the presence of misbehaving, adversarial Byzantines. Each sensor iteratively updates its own estimate based on its previous estimate, its noisy sensor measurement of the parameter, and its neighbors' estimates. A sensor raises a flag to indicate the presence of a Byzantine if any of its neighbors' estimate deviates from its own estimate beyond a given threshold. Under some global observability conditions for the connected normally behaving sensors, it was shown in [3] that if the FRDE algorithm does not detect an attack, then the normally behaving sensors can correctly estimate the target parameter. The false alarm probability of the algorithm may be arbitrarily small with proper selection of parameters, and the approaches to choose these parameters were presented.

References

1. Aurenhammer F (1991) Voronoi diagram-A survey of a fundamental geometric data structure. ACM Comput Surv 23(3):345–405. https://doi.org/10.1145/116873.116880
2. Berkhin P (2002) Survey of clustering data mining techniques. Technical report. Accrue Software Inc, San Jose, CA, USA
3. Chen Y, Kar S, Moura JMF (2018) Resilient distributed estimation through adversary detection. IEEE Trans Signal Process 66(9):2455–2469. https://doi.org/10.1109/TSP.2018.2813330
4. Doucet A, Wang X (2005) Monte Carlo methods for signal processing: a review in the statistical signal processing context. IEEE Signal Process Mag 22(6):152–170. https://doi.org/10.1109/MSP.2005.1550195
5. Fudenberg D, Tirole J (1991) Game theory. MIT Press, Cambridge, MA
6. Kar S, Moura JMF (2013) Consensus+innovations distributed inference over networks. IEEE Signal Process Mag 30(3):99–109. https://doi.org/10.1109/MSP.2012.2235193
7. Li D (2003) Hu YH (2003) Energy-based collaborative source localization using acoustic micro-sensor array. EURASIP J Appl Signal Process 4:331–337. https://doi.org/10.1155/S1110865703212075
8. Masazade E, Niu R, Varshney PK (2010) Energy aware iterative source localization schemes for wireless sensor networks. IEEE Trans Signal Process 58(9):4824–4835. https://doi.org/10.1109/TSP.2010.2051433
9. Meesookho C, Mitra U, Narayanan S (2008) On energy-based acoustic source localization for sensor networks. IEEE Trans Signal Process 56(1):365–377. https://doi.org/10.1109/TSP.2007.900757
10. Niu R, Varshney PK (2006) Target location estimation in sensor networks with quantized data. IEEE Trans Signal Process 54(12):4519–4528. https://doi.org/10.1109/TSP.2006.882082
11. Ozdemir O, Niu R, Varshney PK (2009) Channel aware target localization with quantized data in wireless sensor networks. IEEE Trans Signal Process 57(3):1190–1202. https://doi.org/10.1109/TSP.2008.2009893
12. Rawat A, Anand P, Chen H, Varshney PK (2011) Collaborative spectrum sensing in the presence of Byzantine attacks in cognitive radio networks. IEEE Trans Signal Process 59(2):774–786. https://doi.org/10.1109/TSP.2010.2091277
13. Ribeiro A, Giannakis GB (2006) Bandwidth-constrained distributed estimation for wireless sensor networks- Part I: Gaussian case. IEEE Trans Signal Process 54(3):1131–1143. https://doi.org/10.1109/TSP.2005.863009
14. Shen X, Hu YH (2005) Maximum likelihood multiple-source localization using acoustic energy measurements with wireless sensor network. IEEE Trans Signal Process 53(1):44–53. https://doi.org/10.1109/TSP.2004.838930
15. Van Trees HL (1968) Detection, estimation and modulation theory, vol 1. Wiley, New York, NY
16. Van Trees HL, Bell KL (2007) Bayesian bounds for parameter estimation and nonlinear filtering/tracking. Wiley-IEEE, Hoboken, NJ
17. vanBrunt B (2004) The calculus of variations. Springer, New York, NY
18. Vempaty A, Agrawal K, Chen H, Varshney PK (2011) Adaptive learning of Byzantines' behavior in cooperative spectrum sensing. In: Proceedings of the 2011 IEEE Wireless Communications and Networking Conference (WCNC), pp 1310–1315. https://doi.org/10.1109/WCNC.2011.5779320
19. Vempaty A, Ozdemir O, Agrawal K, Chen H, Varshney PK (2013) Localization in wireless sensor networks: Byzantines and mitigation techniques. IEEE Trans Signal Process 61(6):1495–1508. https://doi.org/10.1109/TSP.2012.2236325
20. Vempaty A, Han YS, Varshney PK (2014) Target localization in wireless sensor networks using error correcting codes. IEEE Trans Inf Theory 60(1):697–712. https://doi.org/10.1109/TIT.2013.2289859

21. Wang TY, Han YS, Varshney PK, Chen PN (2005) Distributed fault-tolerant classification in wireless sensors networks. IEEE J Sel Areas Commun 23(4):724–734. https://doi.org/10.1109/JSAC.2005.843541
22. Wang TY, Han YS, Chen B, Varshney PK (2006) A combined decision fusion and channel coding scheme for distributed fault-tolerant classification in wireless sensors networks. IEEE Trans Wireless Commun 5(7):1695–1705. https://doi.org/10.1109/TWC.2006.1673081
23. Yao C, Chen PN, Wang TY, Han YS, Varshney PK (2007) Performance analysis and code design for minimum Hamming distance fusion in wireless sensor networks. IEEE Trans Inf Theory 53(5):1716–1734. https://doi.org/10.1109/TIT.2007.894670
24. Zhang J, Blum RS, Lu X, Conus D (2015) Asymptotically optimum distributed estimation in the presence of attacks. IEEE Trans Signal Process 63(5):1086–1101. https://doi.org/10.1109/TSP.2014.2386281

Chapter 5
Some Additional Topics on Distributed Inference

While the previous two chapters considered the problems of distributed detection and estimation in the presence of Byzantines and with binary quantized data, this chapter focuses on their generalizations. Specifically, Sect. 5.1 considers the general distributed inference problem in the presence of Byzantines when the sensors use an M-ary quantizer. This problem is addressed from the Byzantine attacker's perspective, and the critical power and optimal attack strategy are derived that are independent of the inference problem. Then, Sect. 5.2 considers the problem of target tracking using quantized data in the presence of Byzantines. Here again, we follow the general flow of addressing the problem first from the attacker's perspective to derive the optimal attack strategy. The problem is then considered from the network's perspective to determine the optimal defense mechanisms.

5.1 Distributed Inference with M-ary Quantized Byzantine Data

In [4], a generalized distributed inference problem was considered in the presence of Byzantines but with M-ary quantization at the local sensors. The authors considered the framework of Byzantine attacks when Byzantine sensors do not have complete knowledge about the true state of the phenomenon of interest (POI). They also consider that the Byzantine attacker is ignorant about the quantization thresholds used at the sensors to generate the M-ary symbols.

5.1.1 System Model

Continuing with our notation, consider a sensor network with N sensors where α fraction of them are Byzantines. The observation x_i at the ith sensor is quantized

A. Vempaty et al., *Secure Networked Inference with Unreliable Data Sources*,
https://doi.org/10.1007/978-981-13-2312-6_5

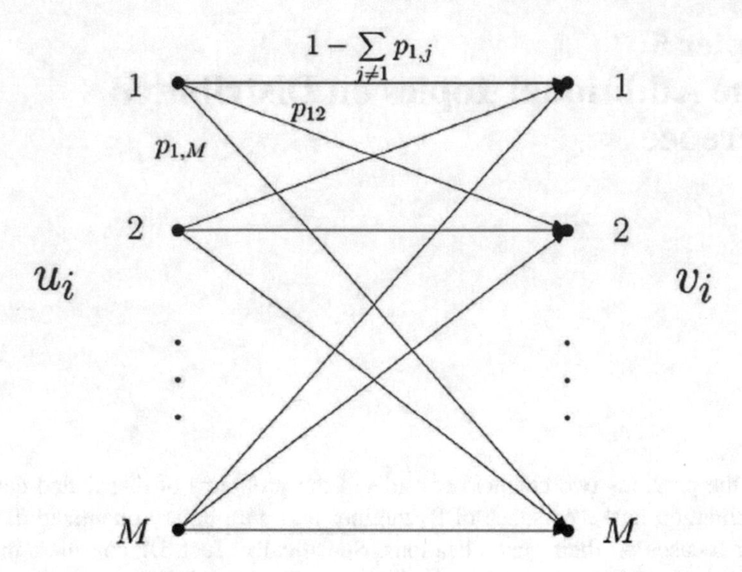

Fig. 5.1 Byzantine attack model [4]

to one of M symbols, $u_i \in \{1, \ldots, M\}$. Since the Byzantine sensors do not transmit their true quantized data, the transmitted symbol v_i at the ith sensor can be different from u_i. If the sensor is honest, then $v_i = u_i$. Otherwise, it is assumed that the Byzantines modify their local symbol $u_i = l$ to $v_i = m$ with a probability p_{lm}, as shown via the attack model depicted in Fig. 5.1. Denote the transition probabilities using a row-stochastic matrix $\mathbb{P} = \{p_{lm}\}$. Since the attacker has no information on the sensor thresholds, $\mathbb{P}$ is independent of sensor observations. The FC performs inference based on the observations $\mathbf{v} = [v_1, \ldots, v_N]$. The general analysis of the inference problem was performed in [4] by abstracting the inference parameter as θ which could be discrete (binary) or continuous depending on the specific inference task.

5.1.2 Analysis from Attacker's Perspective

In [4], the attack strategies were determined by analyzing the system's performance as determined by the probability mass function (pmf) $P(\mathbf{v}|\theta)$. First, observe that the pmf $P(v_i|\theta)$ for any ith sensor can be written in terms of α, $\mathbb{P}$, and $P(u_i|\theta)$ as [4]

$$P(v_i = m|\theta) = [(1-\alpha) + \alpha p_{mm}] + \sum_{l\neq m} \{\alpha p_{lm} - [(1-\alpha) + \alpha p_{mm}]\}\, P(u_i = l|\theta). \tag{5.1}$$

Now note that, as discussed before, the goal of a Byzantine attack is to *blind* the FC with the least amount of effort (in other words, minimum α). The FC is completely blind to the data, when the pmf $P(v_i = m|\theta) = \frac{1}{M}$ for all $1 \leq m \leq M$, and independent of θ. When this happens, the FC is unable to extract any information from the observations $\mathbf{v}$.

From (5.1), observe that $P(v_i = m|\theta)$ consists of two terms. The first one is based on prior knowledge, and the second term conveys information based on the observations. The FC is blind to the observations, if the second term is made *zero*. And since the attacker does not have any knowledge regarding $P(u_i = l|\theta))$, the second term can be made zero by setting

$$\alpha p_{lm} - [(1-\alpha) + \alpha p_{mm}] = 0, \forall l \neq m. \tag{5.2}$$

This results in the conditional probability $P(v_i = m|\theta) = (1-\alpha) + \alpha p_{mm}$ becoming independent of the observations. The FC has to now perform inference only using its prior information about θ.

Based on the condition in (5.2), the attacker's strategy was identified in [4]. Since we need $P(v_i = m|\theta)$ to be equiprobable, we have

$$(1-\alpha) + \alpha p_{mm} = \frac{1}{M} \Longleftrightarrow \alpha = \frac{M-1}{(1-p_{mm})M}. \tag{5.3}$$

To minimize α, one needs to make $p_{mm} = 0$. In other words, $\alpha^* = \frac{M-1}{M}$, and the optimal strategy has $p_{mm} = 0$. Substituting these values in (5.2), we get

$$p_{lm} = \frac{1}{M-1}. \tag{5.4}$$

This result is summarized as follows.

Theorem 5.1 ([4]) *If the Byzantine attacker has no knowledge of the quantization thresholds employed at each sensor, then the optimal Byzantine attack is given as*

$$p_{lm} = \begin{cases} \frac{1}{M-1} & l \neq m \\ 0 & \textit{otherwise}. \end{cases} \tag{5.5}$$

and $\alpha^* = \frac{M-1}{M}$.

Theorem 5.1 was also extended in [4] to the case where the channels between sensors (attackers) are not perfect.

Theorem 5.2 ([4]) *Let the Byzantine attackers have no knowledge about the sensors' quantization thresholds and the FC's channel matrix be* $\mathbb{Q}$. *Let* $\mathbf{0}$, $\mathbf{1}$, $\mathbb{U}$, *and* $\mathbb{I}$, *respectively, be the all-zero vector, all-one vector, all-one matrix, and identity matrix. If* $\mathbb{Q}$ *is non-singular and if* $\mathbf{0} \leq (\mathbb{Q}^T)^{-1}\mathbf{1} \leq M\mathbf{1}$, *then the optimal Byzantine attack is*

$$\mathbb{P} = \frac{1}{\alpha^* M}\mathbb{U}\mathbb{Q}^{-1} - \frac{1-\alpha^*}{\alpha^*}\mathbb{I} \tag{5.6}$$

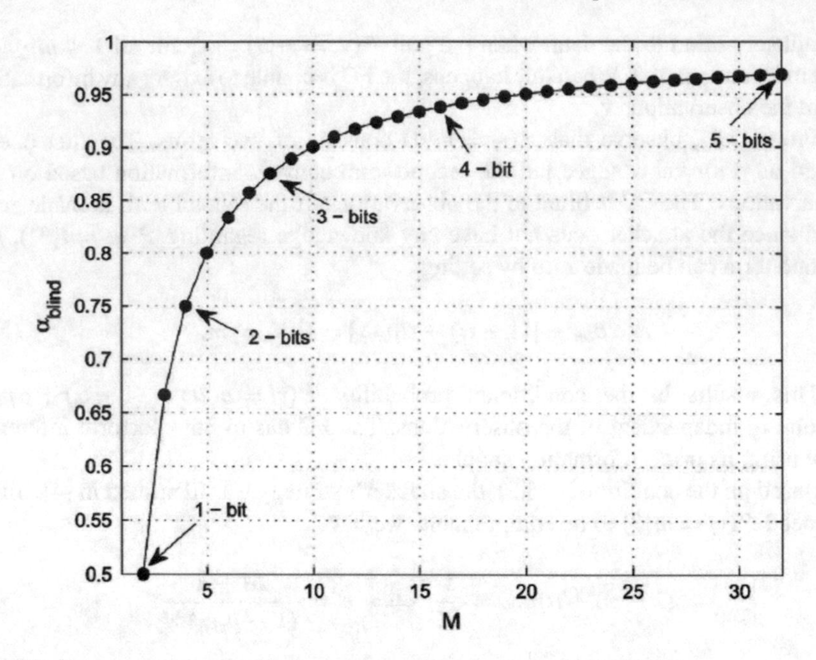

Fig. 5.2 Improvement in α^* with increasing number of quantization levels [4]

Table 5.1 Improvement in α^* with increasing number of quantization bits, $\log_2 M$ [4]

Quantization bits	α^*
1	0.5
2	0.75
3	0.875
4	0.9375
5	0.9688
6	0.9844
7	0.9922
8	0.9961

and

$$\alpha^* = 1 - \frac{1}{M} \max \{(\mathbb{Q}^T)^{-1}\mathbf{1}\}. \tag{5.7}$$

Figure 5.2 and Table 5.1 show how α^* scales with increasing quantization alphabet size M. Observe that an increase of one additional quantization bit (2-bit quantization) in the quantization scheme instead of 1-bit quantization (binary quantization discussed in previous chapters) increases the critical power α^* from 0.5 to 0.75. This trend continues and when the sensors employ an 8-bit quantizer, then the critical power reaches 99.6%.

This improvement in security performance is at the cost of an increased communication cost in terms of energy and bandwidth as the number of quantization bits increases. Therefore, there exists a trade-off between the communication cost and the security guarantees.

5.2 Target Tracking with Quantized Byzantine Data

Target tracking is a generalized version of the estimation problem discussed in Chap. 4. We now present results on the problem of target tracking in the presence of Byzantine data [8].

5.2.1 System Model

Consider a single target moving in a two-dimensional Cartesian coordinate plane whose dynamics is defined by the four-dimensional state vector $\boldsymbol{\theta_k} = [x_{tk}\ y_{tk}\ \dot{x}_{tk}\ \dot{y}_{tk}]^T$ where x_{tk} and y_{tk} denote the x and y coordinates of the target, respectively, at time k. $\dot{x}_{tk}$ and $\dot{y}_{tk}$ denote the first-order derivatives (velocities) in x- and y-directions, respectively. Target motion is defined by the white-noise acceleration model [2] as described below:

$$\boldsymbol{\theta}_{\mathbf{k}} = \mathbf{F}\boldsymbol{\theta}_{\mathbf{k-1}} + \nu_k, \tag{5.8}$$

where

$$\mathbf{F} = \begin{bmatrix} 1 & 0 & T & 0 \\ 0 & 1 & 0 & T \\ 0 & 0 & 1 & 0 \\ 0 & 0 & 0 & 1 \end{bmatrix} \tag{5.9}$$

and ν_k is the additive white Gaussian process noise which is assumed to be zero mean with covariance matrix $\mathbf{Q}$ given by

$$\mathbf{Q} = q \begin{bmatrix} \frac{T^3}{3} & 0 & \frac{T^2}{2} & 0 \\ 0 & \frac{T^3}{3} & 0 & \frac{T^2}{2} \\ \frac{T^2}{2} & 0 & T & 0 \\ 0 & \frac{T^2}{2} & 0 & T \end{bmatrix}, \tag{5.10}$$

where q and T denote the process noise parameter and the time interval between adjacent sensor measurements, respectively. Assume that the FC has an exact knowledge of the target state-space model given in (5.8) and the process noise statistics. There are N sensors deployed in the network. The dynamic target radiates a signal which is assumed to follow an isotropic power attenuation model same as the location estimation problem:

$$a_{ik}^2 = P_0 \left(\frac{d_0}{d_{ik}}\right)^n, \tag{5.11}$$

where a_{ik} is the signal amplitude received at the ith sensor at time instant k, P_0 is the power measured at a reference distance d_0, n is the path-loss exponent, and d_{ik} is the distance between the target and the ith sensor at the kth time step. Without loss of generality, assume $d_0 = 1$ and $n = 2$. The signal amplitude is assumed to be corrupted by additive white Gaussian noise (AWGN) at each sensor:

$$s_{ik} = a_{ik} + n_i, \tag{5.12}$$

where s_{ik} is the corrupted signal at the ith sensor at time instant k and the noise n_i follows $\mathcal{N}(0, \sigma^2)$. Assume this noise to be independent across sensors. Due to energy and bandwidth constraints, each sensor quantizes its received signal s_{ik} locally using a binary quantizer and the quantized data is sent to the FC. Consider threshold quantizers for their simplicity in terms of implementation and analysis:

$$v_{ik} = \begin{cases} 0 & s_{ik} < \eta_{ik} \\ 1 & s_{ik} > \eta_{ik} \end{cases}. \tag{5.13}$$

In (5.13), v_{ik} is the locally quantized binary measurement of the ith sensor and η_{ik} is the quantization threshold used by this sensor at time instant k. The FC receives the binary vector $\mathbf{u_k} = [u_{1k}, \cdots, u_{Nk}]$ from all the sensors in the network, where u_{ik} may not be equal to v_{ik} due to the presence of Byzantine sensors in the network (c.f. Sect. 5.2.1). After collecting $\mathbf{u_k}$, the FC sequentially estimates the target state θ_k using a sequential importance resampling (SIR) particle filter [1].

Performance Metrics

As used before for the location estimation problem, PCRLB is used as the metric for tracking performance. Let $\hat{\boldsymbol{\theta}}_\mathbf{k}(\mathbf{u_{1:k}})$ be an estimator of the state vector $\boldsymbol{\theta}_\mathbf{k}$ at time k, given all the available measurements $\mathbf{u_{1:k}} = [\mathbf{u_1} \cdots \mathbf{u_k}]$ up to time k. Then, the mean square error (MSE) matrix of the estimation error at time k is bounded below by the PCRLB $\mathbf{J_k}^{-1}$ [3],

$$\mathbf{B_k} = E\left\{\left[\hat{\boldsymbol{\theta}}_\mathbf{k}(\mathbf{u}_{1:k}) - \boldsymbol{\theta}_k\right]\left[\hat{\boldsymbol{\theta}}_\mathbf{k}(\mathbf{u}_{1:k}) - \boldsymbol{\theta}_k\right]^T\right\} \geq \mathbf{J_k}^{-1}, \tag{5.14}$$

where $\mathbf{J_k}$ is the Fisher information matrix (FIM). Tichavský *et al.* in [7] provide a recursive approach to calculate this sequential FIM $\mathbf{J_k}$:

$$\mathbf{J_{k+1}} = \mathbf{D_k^{22}} - \mathbf{D_k^{21}}\left(\mathbf{J_k} + \mathbf{D_k^{11}}\right)^{-1}\mathbf{D_k^{12}}, \tag{5.15}$$

where

$$\mathbf{D_k^{11}} = E\left\{-\nabla_{\boldsymbol{\theta}_\mathbf{k}}\nabla_{\boldsymbol{\theta}_\mathbf{k}}^T \log p\left(\boldsymbol{\theta}_\mathbf{k+1}|\boldsymbol{\theta}_\mathbf{k}\right)\right\}, \tag{5.16}$$

$$\mathbf{D_k^{12}} = E\left\{-\nabla_{\boldsymbol{\theta}_\mathbf{k}}\nabla_{\boldsymbol{\theta}_\mathbf{k+1}}^T \log p\left(\boldsymbol{\theta}_\mathbf{k+1}|\boldsymbol{\theta}_\mathbf{k}\right)\right\}, \tag{5.17}$$

$$\mathbf{D_k^{21}} = E\left\{-\nabla_{\boldsymbol{\theta}_\mathbf{k+1}}\nabla_{\boldsymbol{\theta}_\mathbf{k}}^T \log p\left(\boldsymbol{\theta}_\mathbf{k+1}|\boldsymbol{\theta}_\mathbf{k}\right)\right\} = \left(\mathbf{D_k^{12}}\right)^T, \tag{5.18}$$

$$\begin{aligned}\mathbf{D_k^{22}} &= E\left\{-\nabla_{\boldsymbol{\theta}_\mathbf{k+1}}\nabla_{\boldsymbol{\theta}_\mathbf{k+1}}^T \log p\left(\boldsymbol{\theta}_\mathbf{k+1}|\boldsymbol{\theta}_\mathbf{k}\right)\right\} \\ &\quad + E\left\{-\nabla_{\boldsymbol{\theta}_\mathbf{k+1}}\nabla_{\boldsymbol{\theta}_\mathbf{k+1}}^T \log p\left(\mathbf{u}_{k+1}|\boldsymbol{\theta}_\mathbf{k+1}\right)\right\} \\ &= \mathbf{D_k^{22,a}} + \mathbf{D_k^{22,b}}.\end{aligned} \tag{5.19}$$

The derivative operator $\nabla_{\boldsymbol{\theta}_\mathbf{k}}$ in (5.16)–(5.19) is defined as

$$\nabla_{\theta_k} = \left[\frac{\partial}{\partial x_{tk}}, \frac{\partial}{\partial y_{tk}}, \frac{\partial}{\partial \dot{x}_{tk}}, \frac{\partial}{\partial \dot{y}_{tk}}\right]^T \tag{5.20}$$

and the expectations in (5.16)–(5.19) are taken with respect to the joint probability distribution $p\left(\boldsymbol{\theta}_\mathbf{0:k+1}, \mathbf{u}_{1:k+1}\right)$. The a priori probability density function (pdf) of the target state $p_0(\boldsymbol{\theta}_\mathbf{0})$ can be used to calculate the initial FIM as $\mathbf{J_0} = E\left\{-\nabla_{\boldsymbol{\theta}_\mathbf{0}}\nabla_{\boldsymbol{\theta}_\mathbf{0}}^T \log p_0\left(\boldsymbol{\theta}_\mathbf{0}\right)\right\}$.

For the target dynamic model and the measurement model used in this section, the expressions in (5.16)–(5.19) simplify to

$$\mathbf{D_k^{11}} = \mathbf{F}^T\mathbf{Q}^{-1}\mathbf{F}, \tag{5.21}$$

$$\mathbf{D_k^{12}} = \left(\mathbf{D_k^{21}}\right)^T = -\mathbf{F}^T\mathbf{Q}^{-1}, \tag{5.22}$$

$$\mathbf{D_k^{22}} = \mathbf{Q}^{-1} + \mathbf{D_k^{22,b}}. \tag{5.23}$$

Note that $\mathbf{D_k^{22,b}}$ is the only term that depends on the observations $\mathbf{u}_{1:k}$ of the local sensors.

Attack Model

For this target-tracking problem, the focus here is only on independent attacks. Since the quantized observations transmitted by the local sensors are binary in nature, the Byzantines can attack the network by changing this binary quantized data. Let $M = \alpha N$ be the number of Byzantines in the network. It is assumed that the FC knows the fraction of Byzantines (α) but does not the know the identity of the Byzantines. An honest sensor sends its true observation to the FC and therefore, $u_{ik} = v_{ik}$, whereas for the Byzantines, we assume that they flip their quantized binary measurements with probability p. Therefore, under the Gaussian noise assumption, the probability that $u_{ik} = 1$ is given by

$$P(u_{ik} = 1|\boldsymbol{\theta}_\mathbf{k}, i = Honest) = Q\left(\frac{\eta_{ik} - a_{ik}}{\sigma}\right), \tag{5.24}$$

$$P(u_{ik}=1|\boldsymbol{\theta}_{\mathbf{k}}, i=Byzantine) = p\left(1-Q\left(\frac{\eta_{ik}-a_{ik}}{\sigma}\right)\right)+(1-p)Q\left(\frac{\eta_{ik}-a_{ik}}{\sigma}\right), \tag{5.25}$$

where $Q(\cdot)$ is the complementary cumulative distribution function of a standard Gaussian distribution. From the FC's perspective, the probability of a sensor sending '1' is therefore, given by

$$\begin{aligned} P_{ik} &= P(u_{ik}=1|\boldsymbol{\theta}_{\mathbf{k}}) \\ &= (1-\alpha)Q\left(\frac{\eta_{ik}-a_{ik}}{\sigma}\right) \\ &+ \alpha\left(p\left(1-Q\left(\frac{\eta_{ik}-a_{ik}}{\sigma}\right)\right)+(1-p)Q\left(\frac{\eta_{ik}-a_{ik}}{\sigma}\right)\right). \end{aligned} \tag{5.26}$$

Proposition 5.1 ([8]) *Under independent Byzantine attacks, data's contribution* $\mathbf{D}_{\mathbf{k}}^{\mathbf{22,b}}$ *to FIM is given by:*

$$\mathbf{D}_{\mathbf{k}}^{\mathbf{22,b}} = \begin{bmatrix} \mathbf{D}_{\mathbf{k}}^{\mathbf{22,b}}{}_{11} & \mathbf{D}_{\mathbf{k}}^{\mathbf{22,b}}{}_{12} & 0 & 0 \\ \mathbf{D}_{\mathbf{k}}^{\mathbf{22,b}}{}_{21} & \mathbf{D}_{\mathbf{k}}^{\mathbf{22,b}}{}_{22} & 0 & 0 \\ 0 & 0 & 0 & 0 \\ 0 & 0 & 0 & 0 \end{bmatrix}, \tag{5.27}$$

where the nonzero elements are given by

$$\mathbf{D}_{\mathbf{k}}^{\mathbf{22,b}}{}_{11} = \sum_{i=1}^{N} E\left[\frac{(1-2\alpha p)^2 a_{il}^2 e^{-\frac{(\eta_{il}-a_{il})^2}{\sigma^2}}(x_{il}-x_{tl})^2}{2\pi\sigma^2 d_{il}^4 P_{il}(1-P_{il})}\right], \tag{5.28}$$

$$\mathbf{D}_{\mathbf{k}}^{\mathbf{22,b}}{}_{12} = \mathbf{D}_{\mathbf{k}}^{\mathbf{22,b}}{}_{21} = \sum_{i=1}^{N} E\left[\frac{(1-2\alpha p)^2 a_{il}^2 e^{-\frac{(\eta_{il}-a_{il})^2}{\sigma^2}}(x_{il}-x_{tl})(y_{il}-y_{tl})}{2\pi\sigma^2 d_{il}^4 P_{il}(1-P_{il})}\right], \tag{5.29}$$

$$\mathbf{D}_{\mathbf{k}}^{\mathbf{22,b}}{}_{22} = \sum_{i=1}^{N} E\left[\frac{(1-2\alpha p)^2 a_{il}^2 e^{-\frac{(\eta_{il}-a_{il})^2}{\sigma^2}}(y_{il}-y_{tl})^2}{2\pi\sigma^2 d_{il}^4 P_{il}(1-P_{il})}\right]. \tag{5.30}$$

where for $l=k+1$, P_{il} *is given in* (5.26) *and the expectation is taken with respect to* $p\left(\boldsymbol{\theta}_{\mathbf{0:k}}, \mathbf{u}_{1:\mathbf{k}}\right) p\left(\boldsymbol{\theta}_{\mathbf{k+1}}|\boldsymbol{\theta}_{\mathbf{k}}\right)$.

Proof Note that the only nonzero terms of $\mathbf{D}_{\mathbf{k}}^{\mathbf{22,b}}$ are in the 2×2 sub-matrix $\mathbf{D}_{\mathbf{k}}^{\mathbf{22,b}}(1:2, 1:2)$. This is due to the fact that the distribution of data does not depend on the first-order derivatives ($\dot{x}_{tk}$ and $\dot{y}_{tk}$) of the state $\boldsymbol{\theta}_{\mathbf{k}}$. Based on the fact that $\boldsymbol{\theta}_{\mathbf{k}}$, $\boldsymbol{\theta}_{\mathbf{k+1}}$ and $\mathbf{u}_{k+1}$ form a Markov chain, the joint pdf for the expectation can be rewritten as follows

$$p\left(\boldsymbol{\theta}_{\mathbf{0:k+1}}, \mathbf{u}_{1:k+1}\right) = p\left(\boldsymbol{\theta}_{\mathbf{0:k}}, \mathbf{u}_{1:k}\right) p\left(\boldsymbol{\theta}_{\mathbf{k+1}}|\boldsymbol{\theta}_{\mathbf{k}}\right) p\left(\mathbf{u}_{k+1}|\boldsymbol{\theta}_{\mathbf{k+1}}\right). \tag{5.31}$$

The remainder of the proof follows from the definition of $\mathbf{D_k^{22,b}}$ given in (5.19) and using the following

$$\frac{\partial P_{i(k+1)}}{\partial x_{t(k+1)}} = -\frac{(1-2\alpha p)a_{i(k+1)}e^{-\frac{\left(\eta_{i(k+1)}-a_{i(k+1)}\right)^2}{2\sigma^2}}(x_{i(k+1)}-x_{t(k+1)})}{\sigma\sqrt{2\pi}d_{i(k+1)}^2}, \tag{5.32}$$

$$\frac{\partial P_{i(k+1)}}{\partial y_{t(k+1)}} = -\frac{(1-2\alpha p)a_{i(k+1)}e^{-\frac{\left(\eta_{i(k+1)}-a_{i(k+1)}\right)^2}{2\sigma^2}}(y_{i(k+1)}-y_{t(k+1)})}{\sigma\sqrt{2\pi}d_{i(k+1)}^2}. \tag{5.33}$$

□

5.2.2 Analysis from Attacker's Perspective

The goal of Byzantines is naturally to cause as much damage to the functionality of the FC as possible. Following the same terminology as used for the location estimation problem, this event of causing maximum damage is referred to as *blinding* the FC which refers to making the data from the local sensors non-informative for the FC. In this case, this happens when the data's contribution $\mathbf{D_k^{22,b}}$ to FIM approaches zero. In this scenario, the best the fusion center can do is to use the information from the prior knowledge of state transition model to estimate the state. Since PCRLB/FIM is matrix-valued and it is a function of α, we define the blinding condition to mean that the trace of $D_k^{22,b}$ is zero. That is, α^* may be defined as

$$\alpha^* \triangleq \min\{\alpha \,|\, \mathrm{tr}(\mathbf{D_k^{22,b}}(\alpha)) = 0\}. \tag{5.34}$$

For this framework, a closed-form expression for α^* was derived in [8] and further analysis of target tracking in the presence of Byzantines was carried out by carefully observing the expressions derived in Proposition 5.1. Observe that the nonzero elements of $\mathbf{D_k^{22,b}}$, data's contribution to FIM, given by (5.28)–(5.30) all become zero when $\alpha p = \frac{1}{2}$, i.e., $\mathbf{D_k^{22,b}} = \mathbf{0}$ when $\alpha p = \frac{1}{2}$. This implies that the higher the probability of flipping of Byzantines (p), the lower the fraction of Byzantines required to blind the FC from local sensors' data. Therefore, the minimum fraction of Byzantines, α_{blind}, is $\frac{1}{2}$ corresponding to $p = 1$. Byzantines need to be at least 50% in number and should flip their quantized local observations with probability '1' to blind the network.

5.2.3 *Analysis from Network Designer's Perspective*

Now consider the problem from the network designer's perspective. The non-identical threshold design scheme proposed for the location estimation problem in Chap. 4 can also be used for the target-tracking problem. The estimation model here is sequential and described as follows. At time $k = 0$, the local sensors send quantized data using the optimal identical quantizer thresholds proposed in [5]. The FC estimates the target state at every time instant using particle filtering and broadcasts this estimate information to the local sensors. At every subsequent iteration k, all the N sensors send their one-bit quantized observations using their updated local thresholds. In this new scheme, local sensors update their quantizers based on the feedback information (state estimate) they receive from the FC. In order to develop insights, we follow a similar methodology as done for location estimation problem and investigate the case where there are no Byzantines and all the sensors are honest.

Honest Sensors Only

Note that the recursive Fisher information is a function of sensor thresholds from time 1 to time k. Using the target dynamic model in (5.8) and the sensor measurement model in (5.12), one can design optimal static threshold values offline by minimizing a cost function, which could be the trace or the determinant of the recursive PCRLB in (5.15), over the sensor thresholds. However, as was previously shown in [6], the performance can be improved using a dynamic optimal quantizer design. For the dynamic quantizers, quantizer design must not only be based on the system model but must also exploit the feedback mechanism from the FC to the sensors. In order to dynamically design thresholds in real time, one needs to take into account the information contained in the measurements up to time k, i.e., $\mathbf{u_{1:k}}$. It was also shown in [6] that this optimization is complicated and researchers have found ways to simplify the optimization problem. In [8], this problem was simplified by making every local sensor threshold a function of signal amplitude at the sensor and assuming that all the sensors follow the same functional dependence, i.e.,

$$\eta_{ik} = \eta(a_{ik}), \tag{5.35}$$

where $\eta(\cdot)$ is some function and a_{ik} is the amplitude of the observation at the ith sensor at time k. Since the value of a_{ik} is not known, update each threshold iteratively as

$$\eta_{i(k+1)} = \eta(\hat{a}_{ik}), \tag{5.36}$$

where $\hat{a}_{ik}$ is the estimated amplitude at the previous time instant which is estimated by using the state estimate of the kth iteration, $\hat{\boldsymbol{\theta}}_{\mathbf{k}}$. The minimization problem now becomes a variational minimization problem

$$\min_{\eta(\cdot)} \operatorname{tr}\left(\mathbf{D}_{\mathbf{k}}^{22,\mathbf{b}}(\eta(\cdot))\right). \tag{5.37}$$

This is still a difficult problem as the objective function depends on the target's true state $\boldsymbol{\theta}_{\mathbf{k+1}}$ which is an unknown. Therefore, [8] used the heuristic approach similar to the one used for the location estimation problem in Chap. 4 to design the quantizers. This gives the threshold value as $\eta_{i(k+1)} = \eta(\hat{a}_{ik}) = \hat{a}_{ik}$ which means that the threshold of the ith sensor at time $(k+1)$ is the estimated amplitude at this sensor at the previous time instant k. This previous time instant's state estimate $\hat{\theta}_k$ is needed to estimate the amplitude, which is broadcast by the FC to the local sensors.

The analysis when there are Byzantines in the network is similar to the game-theoretical problem discussed in Chap. 4. This problem consists of two players and can be modeled as a zero-sum game. The two players, the FC and the Byzantines, have opposing objectives where the objective of the FC is to maximize the posterior Fisher information $G[\eta^H(\cdot), \eta^B(\cdot)]$, whereas the objective of the Byzantine sensors is to minimize $G[\eta^H(\cdot), \eta^B(\cdot)]$. This problem has been solved by examining the expression of G [8]. When sensors behave independently, the Fisher information (FI) given by Eq. 4.53 in Chap. 4 and in [8] is maximized when honest sensors set their thresholds as $\eta^H(\hat{a}) = \hat{a}$ regardless of the value of η^B. For $\alpha p \leq \frac{1}{2}$, the Byzantines, who try to minimize this posterior FI, achieve the minimization similarly by setting $\eta^B(\hat{a}) = \hat{a}$ regardless of the value of η^H. This result is expected as the Byzantines flip their observations with a probability p. Observe that if the Byzantines use this thresholding scheme, the probability of a Byzantine sending a '1' is

$$P(u=1|a, Byzantine) = p\left(1 - Q\left(\frac{\eta^B(\hat{a}) - a}{\sigma}\right)\right) + (1-p)Q\left(\frac{\eta^B(\hat{a}) - a}{\sigma}\right) \tag{5.38}$$

which becomes $\frac{1}{2}$, when $\eta^B(\hat{a}) \approx a$, irrespective of the value of p. Also, since $\eta^H(\hat{a}) \approx a$, it implies that the honest sensors also send $0/1$ with probability approximately equal to $\frac{1}{2}$.

Simulation Results

Simulation results presented show the effectiveness of the dynamic non-identical threshold scheme in the presence of Byzantines for the target-tracking problem. Consider a network of N sensors deployed in a grid over a $200\,m \times 200\,m$ area. The sensor density, defined as the number of sensors per unit area, is denoted by ρ. The target is assumed to emit power $P_0 = 25000$, and the local observations at the sensors are assumed to be corrupted by AWGN with zero mean and variance $\sigma^2 = 1$. The target state dynamics is modeled as follows: The initial state distribution is assumed to be Gaussian with mean $\boldsymbol{\mu_0} = [-80\ -80\ 2\ 2]^T$ and covariance $\Sigma_0 = \text{diag}[10^2\ 10^2\ .5^2\ .5^2]$, the target motion model is assumed to be a near constant velocity model, and the process noise parameter is $q = 0.16$. The total observation time duration is 60s, and it is assumed that the observations are made every $T = 1$s. For particle filtering, we use $N_p = 1000$ particles. As we have shown that the optimal strategy for the Byzantines is to flip their local observations with probability '1' for which the $\alpha_{blind} = 0.5$, we consider the case when the fraction of Byzantines $\alpha \leq 0.5$.

The identical threshold scheme uses constant thresholds $\eta_{ik} = 1.7$ for $i = 1, \cdots, N$ and $k = 1, \cdots, 60$ which is the optimal static threshold designed

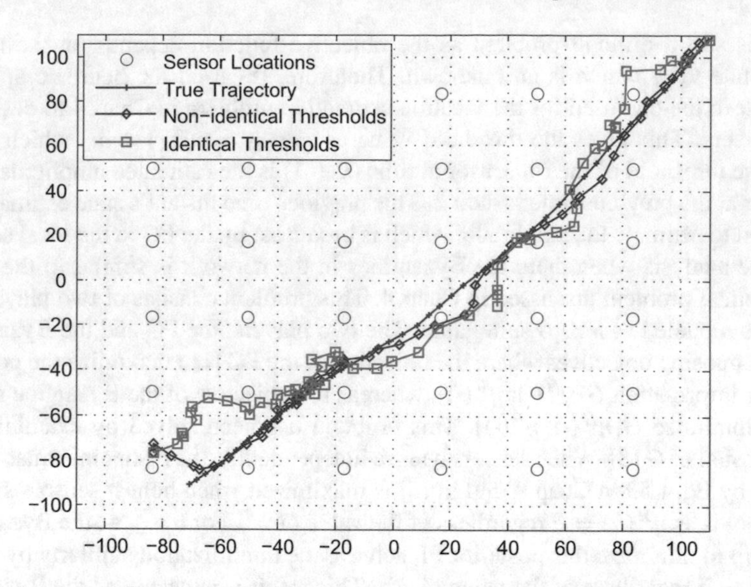

Fig. 5.3 Example of tracking using the two schemes when $\alpha = 0.1$ [8]

in [5]. The dynamic non-identical threshold scheme follows the update mechanism proposed in this section. Figures 5.3 and 5.4 show the improvement in tracking performance when the non-identical threshold scheme is used instead of the identical threshold scheme. For $\alpha = 0.1$, Fig. 5.3 shows the estimated tracks and the true track for a particular realization. It can be observed that the estimation error for both the identical and non-identical threshold schemes is not very different. However, when $\alpha = 0.3$, as seen in Fig. 5.4, the error is significantly reduced when the non-identical threshold scheme is used over the identical threshold scheme.

Now, let us compare the average performance of both the schemes characterized by the average root mean square error (RMSE) over 100 Monte Carlo runs. Figure 5.5 shows this comparison of the two schemes for two different values of N: $N = 36$ ($\rho = 9 \times 10^{-4}$) and $N = 100$ ($\rho = 25 \times 10^{-4}$). As Fig. 5.5 shows, the proposed scheme performs better and it is more robust than the identical threshold scheme in the presence of Byzantines. However, note that both the schemes have the same performance when $\alpha = \alpha_{blind} = 0.5$, since $\alpha = 0.5$ means that the FC is blind to sensor data.

5.3 Summary

In this chapter, we first presented results from [4] which addressed the general distributed inference problem using M-ary quantization. By considering the problem from the attacker's perspective, the optimal Byzantine attack strategy was

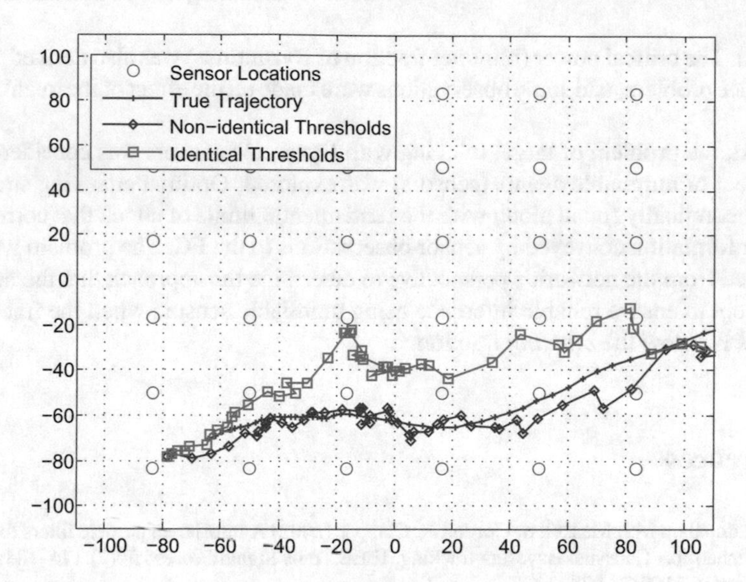

Fig. 5.4 Example of tracking using the two schemes when $\alpha = 0.3$ [8]

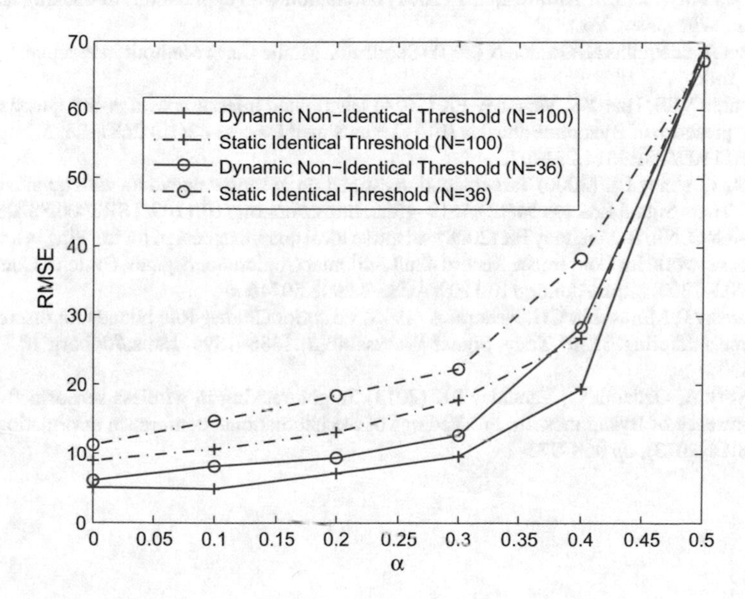

Fig. 5.5 Tracking performance comparison of the two schemes in the presence of Byzantines in terms of RMSE for different values of N [8]

derived. The critical power (blinding fraction of Byzantines) was also derived for this inference problem, and some observations were made on the effect of the quantization levels.

Next, the problem of target tracking with Byzantine sensors was considered and the effect of unreliable agents (sensors) was explored. Optimal attacking strategies were theoretically found along with the fundamental limits of attack that correspond to no information conveyed by sensor observations to the FC. The problem was also analyzed from the network's perspective to determine the approach that the network can adopt to ensure reliable inference using unreliable sensors when the fraction of sensors is below the *blinding* fraction.

References

1. Arulampalam MS, Maskell S, Gordon N, Clapp T (2002) A tutorial on particle filters for online nonlinear/non-Gaussian Bayesian tracking. IEEE Trans Signal Process 50(2):174–188. https://doi.org/10.1109/78.978374
2. Bar-Shalom Y, Li XR, Kirubarajan T (2001) Estimation with applications to tracking and navigation. Wiley, New York
3. Doucet A, de Freitas N, Gordon N (2001) Sequential Monte Carlo Methods in Practice. Springer, New York
4. Nadendla VSS, Han YS, Varshney PK (2014) Distributed inference with m-ary quantized data in the presence of Byzantine attacks. IEEE Trans Signal Process 62(10):2681–2695. https://doi.org/10.1109/TSP.2014.2314072
5. Niu R, Varshney PK (2006) Target location estimation in sensor networks with quantized data. IEEE Trans Signal Process 54(12):4519–4528. https://doi.org/10.1109/TSP.2006.882082
6. Ozdemir O, Niu R, Varshney PK (2008) Adaptive local quantizer design for tracking in a wireless sensor network. In: Conference Record 42nd Asilomar Conference Signals, Systems Computers, pp 1202–1206. https://doi.org/10.1109/ACSSC.2008.5074606
7. Tichavský P, Muravchik CH, Nehorai A (1998) Posterior Cramér-Rao bounds for discrete-time nonlinear filtering. IEEE Trans Signal Process 46(2):1386–1395. https://doi.org/10.1109/78.668800
8. Vempaty A, Ozdemir O, Varshney PK (2013) Target tracking in wireless sensor networks in the presence of Byzantines. In: Proceedings of the international conference information fusion (FUSION2013), pp 968–973

Chapter 6
Distributed Inference with Unreliable Data: Some Unconventional Directions

In this chapter, we present some recent research in the area of secured networked inference. These problems deal with unreliable data in distributed inference networks but consider the problem with applications to unusual domains employing innovative approaches. In Sect. 6.1, we discuss the problem of using Byzantines as a counter-attack by the network against an eavesdropper in a distributed inference network. This is done by intentionally injecting corrupted data into the network to deceive the eavesdropper. In Sect. 6.2, we discuss a crowdsourcing setup where multiple human workers perform a distributed inference task. This again is a problem that deals with distributed inference networks where data is unreliable due to human limitations rather than the presence of Byzantine sensors. And finally in Sect. 6.3, a recent paradigm of Human–Machine Inference Networks (HuMaINs) is discussed, which are networked inferencing systems that deal with unreliable data due to the presence of boundedly rational humans and error-prone physical sensors.

6.1 Friendly Byzantines to Improve Secrecy

Previous sections highlighted the negative impact of unreliable data or data falsification on the inference performance of the system. However, it is possible for a system designer to utilize the corrupted data for network's benefit. Motivated by this fact, we study the positive use of falsified data to improve the secrecy performance of a distributed inference system in the presence of an eavesdropper. The secrecy of a detection system against eavesdropping attacks is of utmost importance [7] in many applications. To solve an inference problem where some prior information about the signal is available, a customized measurement scheme could be implemented such that the optimal inference performance is achieved for the particular signal. As an example, for a signal detection problem where the signal of interest is known, the optimal design is the matched filter which is dependent on the signal itself. However,

A. Vempaty et al., *Secure Networked Inference with Unreliable Data Sources*,
https://doi.org/10.1007/978-981-13-2312-6_6

it is possible that the signal that we wish to infer about may evolve over time. Thus, we are often interested in universal or agnostic design.

To achieve this, the authors in [9] proposed a collaborative compressive detection (CCD) framework. The CCD framework comprises a group of spatially distributed nodes which acquire high-dimensional vector observations regarding the phenomenon of interest. Nodes send a compressed summary of their observations to the fusion center (FC) where a global decision is made. These techniques are universal and agnostic to the signal structure and, therefore, are attractive in many practical applications. Furthermore, [9] proposed to use cooperating trustworthy nodes that assist the FC by injecting corrupted data to deceive the eavesdroppers to improve the secrecy performance of the system.

6.1.1 Collaborative Compressive Detection

System Model

Consider two hypotheses H_0 (signal is absent) and H_1 (signal is present). Also, consider a parallel network (see Fig. 6.1), comprised of the FC and a set of N nodes, which faces the task of determining which of the two hypotheses is true. Prior probabilities of the two hypotheses H_0 and H_1 are denoted by P_0 and P_1, respectively. The nodes observe the phenomenon, carry out local compression (low-dimensional random projection), and then send their local summary statistic to the FC. The FC makes a final decision after processing the locally compressed observations. For the ith node observed signal, u_i can be modeled as

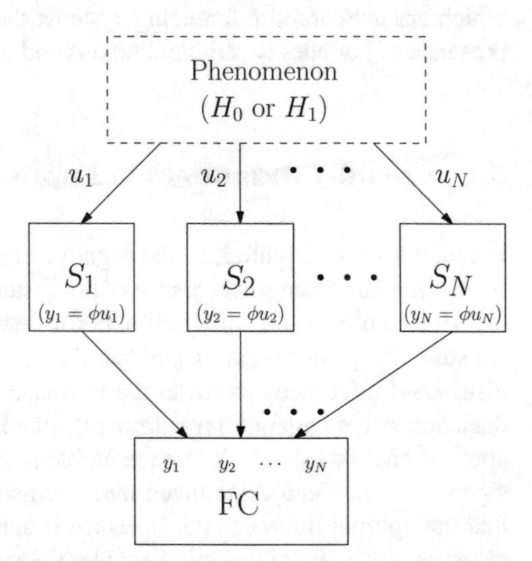

Fig. 6.1 Collaborative compressive detection network [9]

$$\begin{aligned} H_0 &: \quad u_i = v_i \\ H_1 &: \quad u_i = s + v_i, \end{aligned} \tag{6.1}$$

where u_i is the $P \times 1$ observation vector, s is the signal vector (not necessarily sparse) to be detected, v_i is the additive noise distributed as Gaussian $v_i \sim \mathcal{N}(0, \sigma^2 I_P)$ where I_P is the $P \times P$ identity matrix. Observations at the nodes are assumed to be conditionally independent and identically distributed.

Next, each node sends an M-length compressed version y_i of its P-length observation u_i to the FC. The collection of M-length ($< P$) universally sampled observations is given by, $y_i = \phi u_i$, where ϕ is an $M \times P$ random projection operator, which is assumed to be the same for all the nodes, and y_i is the $M \times 1$ compressive observation vector (local summary statistic). Under the two hypotheses, the local summary statistic is

$$\begin{aligned} H_0 &: \quad y_i = \phi v_i \\ H_1 &: \quad y_i = \phi s + \phi v_i. \end{aligned} \tag{6.2}$$

The FC receives compressed observation vectors, $\mathbf{y} = [y_1, \cdots, y_N]$, from the nodes via error-free communication channels and makes the global decision about the phenomenon.

Binary Hypothesis Testing at the Fusion Center

Consider a Bayesian detection problem where the performance criterion at the FC is the probability of error. The FC makes the global decision about the phenomenon by considering the likelihood-ratio test (LRT) which is given by

$$\prod_{i=1}^{N} \frac{f_1(y_i)}{f_0(y_i)} \underset{H_0}{\overset{H_1}{\gtrless}} \frac{P_0}{P_1}. \tag{6.3}$$

Notice that, under the two hypotheses we have the following probability density functions;

$$f_0(y_i) = \frac{\exp(-\frac{1}{2} y_i^T (\sigma^2 \phi \phi^T)^{-1} y_i)}{|\sigma^2 \phi \phi^T|^{1/2} (2\pi)^{M/2}}, \tag{6.4}$$

$$f_1(y_i) = \frac{\exp(-\frac{1}{2} (y_i - \phi s)^T (\sigma^2 \phi \phi^T)^{-1} (y_i - \phi s))}{|\sigma^2 \phi \phi^T|^{1/2} (2\pi)^{M/2}}. \tag{6.5}$$

Taking logarithms on both sides of (6.3), we obtain an equivalent test that simplifies to

$$\sum_{i=1}^{N} y_i^T (\phi \phi^T)^{-1} \phi s \underset{H_0}{\overset{H_1}{\gtrless}} \sigma^2 \log \frac{P_0}{P_1} + \frac{N}{2} (\phi s)^T (\phi \phi^T)^{-1} (\phi s). \tag{6.6}$$

For simplicity, we assume that $P_0 = P_1$. The above test can be written in a compact form as

$$\sum_{i=1}^{N} y_i^T(\phi\phi^T)^{-1}\phi s \underset{H_0}{\overset{H_1}{\gtrless}} \gamma, \tag{6.7}$$

with $\gamma = \frac{N}{2}s^T\phi^T(\phi\phi^T)^{-1}\phi s$. The decision statistic for the collaborative compressive detector is given as

$$\Lambda(\mathbf{y}) = \sum_{i=1}^{N} y_i^T(\phi\phi^T)^{-1}\phi s. \tag{6.8}$$

Performance Analysis of Collaborative Compressive Detection

The authors in [9] analytically characterized the performance of the collaborative compressive detector in terms of the probability of error of the system. The probability of error is defined as

$$P_E = \frac{1}{2}P_F + \frac{1}{2}(1 - P_D), \tag{6.9}$$

where $P_F = P(\Lambda(\mathbf{y}) > \gamma | H_0)$ and $P_D = P(\Lambda(\mathbf{y}) > \gamma | H_1)$ are the probability of false alarm and the probability of detection, respectively. To simplify the notations, let us define

$$\hat{P} = \phi^T(\phi\phi^T)^{-1}\phi \tag{6.10}$$

as the orthogonal projection operator onto row space of ϕ. Using this notation, it is easy to show that

$$\Lambda(\mathbf{y}) \sim \begin{cases} \mathcal{N}(0, \sigma^2 N\|\hat{P}s\|_2^2), \text{ under } H_0 \\ \mathcal{N}(N\|\hat{P}s\|_2^2, \sigma^2 N\|\hat{P}s\|_2^2) \text{ under } H_1 \end{cases} \tag{6.11}$$

where $\|\hat{P}s\|_2^2 = s^T\phi^T(\phi\phi^T)^{-1}\phi s$.

Thus, we have

$$P_F = Q\left(\frac{\frac{N}{2}\|\hat{P}s\|_2^2}{\sigma\sqrt{N}\|\hat{P}s\|_2}\right) = Q\left(\frac{\sqrt{N}}{2\sigma}\|\hat{P}s\|_2\right) \tag{6.12}$$

and

$$P_D = Q\left(\frac{\frac{N}{2}\|\hat{P}s\|_2^2 - N\|\hat{P}s\|_2^2}{\sigma\sqrt{N}\|\hat{P}s\|_2}\right) = Q\left(-\frac{\sqrt{N}}{2\sigma}\|\hat{P}s\|_2\right) \tag{6.13}$$

where $Q(x) = \frac{1}{\sqrt{2\pi}}\int_x^{\infty}\exp(-\frac{u^2}{2})\,du$.

Now, the probability of error can be calculated to be

$$P_E = Q\left(\frac{\sqrt{N}}{2\sigma}\|\hat{P}s\|_2\right). \tag{6.14}$$

Next, let us calculate the deflection coefficient [10] of the system and show its monotonic relationship with the probability of error. Notice that, y_i is distributed under the hypothesis H_j as, $y_i \sim \mathcal{N}(\mu_j^i, \Sigma_j^i)$. The deflection coefficient $D(\mathbf{y})$ can be calculated to be

$$D(\mathbf{y}) = \sum_{i=1}^{N}(\mu_1^i - \mu_0^i)^T(\Sigma_0^i)^{-1}(\mu_1^i - \mu_0^i) \tag{6.15}$$

$$= N\frac{\|\hat{P}s\|_2^2}{\sigma^2}. \tag{6.16}$$

Monotonic relationship between P_E and $D(\mathbf{y})$ can be observed by noticing that

$$P_E = Q\left(\frac{\sqrt{D(\mathbf{y})}}{2}\right). \tag{6.17}$$

Notice that the detection performance of the system is a function of the projection operator $\hat{P}$. In general, this performance could be either quite good or quite poor depending on the random projection matrix ϕ. Next, the authors in [9] provided bounds on the performance of the collaborative compressive detector using the concept of ε-stable embedding.

Stable Embedding: Let $\varepsilon \in (0, 1)$ and $\mathcal{S}, \mathcal{X} \subset \mathbb{R}^P$. We say a mapping ψ is an ε-stable embedding of $(\mathcal{S}, \mathcal{X})$ if

$$(1-\varepsilon)\ \|s-x\|_2^2 \leq \|\psi s - \psi x\|_2^2 \leq (1+\varepsilon)\ \|s-x\|_2^2, \tag{6.18}$$

for all $s \in \mathcal{S}$ and $x \in \mathcal{X}$.

Using this concept, the result can be stated as follows.

Theorem 6.1 ([9]) *Suppose that $\sqrt{\frac{P}{M}}\hat{P}$ provides an ε-stable embedding of $(\mathcal{S}, \{0\})$. Then for any $s \in S$, the probability of error of the collaborative compressive detector satisfies*

$$Q\left(\sqrt{1+\varepsilon}\frac{\sqrt{N}}{2}\sqrt{\frac{M}{P}}\frac{\|s\|_2}{\sigma}\right) \leq P_E \leq Q\left(\sqrt{1-\varepsilon}\frac{\sqrt{N}}{2}\sqrt{\frac{M}{P}}\frac{\|s\|_2}{\sigma}\right).$$

The above expression tells us how much information we lose by using random projections rather than the signal samples themselves. It also tells us how many nodes are needed to collaborate to compensate for the loss due to compression. More specifically, if $N \geq P/M$, the loss due to compression can be recovered. Notice that, for a fixed M as the number of collaborating nodes approaches infinity, i.e., $N \rightarrow \infty$, the probability of error vanishes. On the other hand, to guarantee $P_E \leq \beta$, parameters $M,\ P$, and N should satisfy

$$\frac{NM}{P} \geq \frac{4}{SNR}(Q^{-1}(\beta))^2$$

where $SNR = \frac{\|s\|_2^2}{\sigma^2}$.

In order to more clearly illustrate the behavior of P_E, the following corollary of Theorem 6.1 is established.

Corollary 6.1 ([9]) *Suppose that* $\sqrt{\frac{P}{M}}\hat{P}$ *provides a* ε*-stable embedding of* $(\mathcal{S}, \{0\})$. *Then for any* $s \in S$, *we have*

$$P_E \leq \frac{1}{2}\exp\left(-\frac{1}{4}\frac{NM}{P}\frac{\|s\|_2^2}{\sigma^2}\right).$$

Corollary 6.1 suggests that the error probability vanishes exponentially fast as we increase either the number of measurements M or the number of collaborating nodes N.

Next, let us consider the problem where the network operates in the presence of an eavesdropper who wants to discover the state of the nature being monitored by the system. The FC's goal is to implement the appropriate countermeasures to keep the data regarding the presence of the phenomenon secret from the eavesdropper. More specifically, the goal is to achieve perfect secrecy at the physical layer, by rendering the data available at the eavesdropper to be non-informative.

6.1.2 *Collaborative Compressive Detection in the Presence of an Eavesdropper*

In the collaborative compressive detection framework, the FC receives compressive observation vectors, $\mathbf{y} = [y_1, \cdots, y_N]$, from the nodes and makes the global decision about the phenomenon. The transmission of the nodes, however, may be observed by an eavesdropper who also wants to discover the state of the phenomenon (see Fig. 6.2). To keep the data regarding the presence of the phenomenon secret from the eavesdropper, the authors in [9] proposed to use cooperating trustworthy nodes that assist the FC by providing falsified data to the eavesdroppers to improve the security performance of the system. It was assumed that B out of N nodes (or α fraction of the nodes) falsify their data according to the falsification model explained below.

Nodes tamper with their data y_i and send $\tilde{y}_i$ in the following manner:

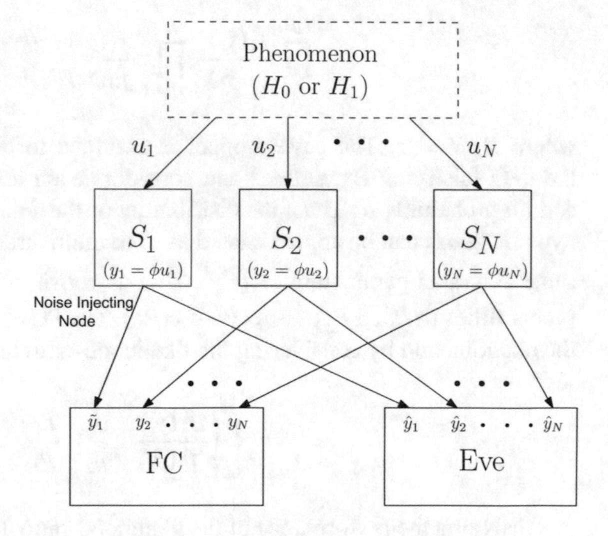

Fig. 6.2 Collaborative compressive detection network in the presence of an eavesdropper [9]

Under H_0:

$$\tilde{y}_i = \begin{cases} \phi(v_i + D_i) \text{ with probability } P_1^0 \\ \phi(v_i - D_i) \text{ with probability } P_2^0 \\ \phi v_i \text{ with probability } (1 - P_1^0 - P_2^0). \end{cases} \tag{6.19}$$

Under H_1:

$$\tilde{y}_i = \begin{cases} \phi(s + v_i + D_i) \text{ with probability } P_1^1 \\ \phi(s + v_i - D_i) \text{ with probability } P_2^1 \\ \phi(s + v_i) \text{ with probability } (1 - P_1^1 - P_2^1). \end{cases} \tag{6.20}$$

where $D_i = \gamma s$ is a $P \times 1$ vector with constant values. The parameter $\gamma > 0$ represents the falsification strength, which is zero for non data falsifying nodes. Assume that the observation model and data falsification parameters are known to both the FC and the eavesdropper. The only information unavailable at the eavesdropper is the identity of the data falsifying Byzantines and considers each node i to be Byzantine with a certain probability α. The FC can distinguish between y_i and $\tilde{y}_i$. Notice that, $\tilde{y}_i$ is distributed under the hypothesis H_0 as a multivariate Gaussian mixture $\mathcal{N}(P_k^0, \tilde{\mu_0}^i, \tilde{\Sigma_0}^i)$ which comes from $\mathcal{N}(\phi D_i, \sigma^2 \phi\phi^T)$ with probability P_1^0, from $\mathcal{N}(-\phi D_i, \sigma^2 \phi\phi^T)$ with probability P_2^0 and from $\mathcal{N}(0, \sigma^2 \phi\phi^T)$ with probability $(1 - P_1^0 - P_2^0)$. Similarly, under the hypothesis H_1 it is distributed as multivariate Gaussian mixture $\mathcal{N}(P_k^1, \tilde{\mu_1}^i, \tilde{\Sigma_1}^i)$ which comes from $\mathcal{N}(\phi(s + D_i), \sigma^2 \phi\phi^T)$ with probability P_1^1, from $\mathcal{N}(\phi(s - D_i), \sigma^2 \phi\phi^T)$ with probability P_2^1 and from $\mathcal{N}(\phi s, \sigma^2 \phi\phi^T)$ with probability $(1 - P_1^1 - P_2^1)$. The FC makes the global decision about the phenomenon by considering the likelihood-ratio test (LRT) which is given by

$$\prod_{i=1}^{B} \frac{f_1(\tilde{y}_i)}{f_0(\tilde{y}_i)} \prod_{i=B+1}^{N} \frac{f_1(y_i)}{f_0(y_i)} \underset{H_0}{\overset{H_1}{\gtrless}} \frac{P_0}{P_1}, \tag{6.21}$$

where $B/N = \alpha$. The eavesdropper is assumed to be unaware of the identity of the data falsifying Byzantines and considers each node i to be Byzantine with a certain probability α. Thus, the distribution of the data $\hat{y}_i$ at the eavesdropper under hypothesis H_j can be approximated as i.i.d. multivariate Gaussian mixture with the same Gaussian parameters $\mathcal{N}(\tilde{\mu}_j{}^i, \tilde{\Sigma}_j{}^i)$ as above, however, with rescaled mixing probabilities $(\alpha P_1^j, \alpha P_2^j, 1 - \alpha P_1^j - \alpha P_2^j)$. The FC makes the global decision about the phenomenon by considering the likelihood-ratio test (LRT) which is given by

$$\prod_{i=1}^{N} \frac{f_1(\hat{y}_i)}{f_0(\hat{y}_i)} \underset{H_0}{\overset{H_1}{\gtrless}} \frac{P_0}{P_1}. \tag{6.22}$$

Analyzing the performance of the likelihood-ratio detector in (6.21) and (6.22) in closed form is difficult in general. Therefore, the deflection coefficient was used in [9] in lieu of the probability of error of the system. Deflection coefficient is closely related to the output signal to noise ratio and widely used in optimizing the performance of detection systems. As stated earlier, the deflection coefficient is defined as

$$D(y_i) = (\mu_1^i - \mu_0^i)^T (\Sigma_0^i)^{-1} (\mu_1^i - \mu_0^i), \tag{6.23}$$

where μ_j^i and Σ_j^i are the mean and the covariance matrix of y_i under the hypothesis H_j, respectively. Using these notations, the deflection coefficient at the FC can be written as

$$D(FC) = BD(\tilde{y}_i) + (N - B)D(y_i). \tag{6.24}$$

Dividing both sides of the above equation by N, we get

$$D_{FC} = \alpha D(\tilde{y}_i) + (1 - \alpha) D(y_i), \tag{6.25}$$

where $D_{FC} = \frac{D(FC)}{N}$ and will be used as an equivalent performance metric. Similarly, the deflection coefficient at the eavesdropper can be written as

$$D_{EV} = \frac{D(EV)}{N} = D(\hat{y}_i). \tag{6.26}$$

Notice that both D_{FC} and D_{EV} are functions of the compression ratio c and data falsification parameters (α, γ) which are under the control of the FC. This motivated the authors in [9] to design the optimal system parameters under a physical layer secrecy constraint. The problem can be formally stated as:

$$\begin{aligned} \underset{c,\alpha,\gamma}{\text{maximize}} \quad & \alpha D(\tilde{y}_i) + (1-\alpha) D(y_i) \\ \text{subject to} \quad & D(\hat{y}_i) \leq \tau \end{aligned} \tag{P6.1}$$

where $c = M/P$ is the compression ratio. We refer to $D(\hat{y}_i) \leq \tau$, where $\tau \geq 0$, as the physical layer secrecy constraint which reflects the security performance of the system. The case where $\tau = 0$, or equivalently $D(\hat{y}_i) = 0$, is referred to as the perfect secrecy constraint. In the wiretap channel literature, it is typical to consider the maximum degree of information achieved by the main user (FC), while the information of the eavesdropper is exactly zero. This is commonly referred to as perfect secrecy regime [40].

A Closed-Form Expression of the Deflection Coefficient at the FC

The deflection coefficient at the FC is

$$D_{FC} = \alpha D(\tilde{y}_i) + (1-\alpha) D(y_i), \tag{6.27}$$

where $D(y_i) = \|\hat{P}s\|_2^2/\sigma^2$. Next, to calculate $D(\tilde{y}_i)$, observe that $\tilde{y}_i$ is distributed as a multivariate Gaussian mixture with

$$\tilde{\mu}_0^i = (P_1^0 - P_2^0)\phi D_i \tag{6.28}$$

$$\tilde{\mu}_1^i = (P_1^1 - P_2^1)\phi D_i + \phi s \tag{6.29}$$

$$\tilde{\Sigma}_0^i = \sigma^2 \phi\phi^T + \sum_{j=1}^{3} P_j^0 (\tilde{\mu}_0^i(j) - \tilde{\mu}_0^i)(\tilde{\mu}_0^i(j) - \tilde{\mu}_0^i)^T, \tag{6.30}$$

where $\tilde{\mu}_0^i(1) = \phi D_i$, $\tilde{\mu}_0^i(2) = -\phi D_i$, $\tilde{\mu}_0^i(3) = 0$ and $P_3^0 = 1 - P_1^0 - P_2^0$. After some calculation, it can be shown that

$$\tilde{\Sigma}_0^i = \sigma^2 \phi\phi^T + P_t[\phi D_i D_i^T \phi^T], \tag{6.31}$$

where $P_t = P_1^0 + P_2^0 + (P_1^0 - P_2^0)^2$. Also, notice that $\tilde{\Sigma}_0^i$ is of the form $A + bb^T$. Now, using Sherman–Morrison formula [27], its inverse can be calculated to be

$$(\tilde{\Sigma}_0^i)^{-1} = \frac{(\phi\phi^T)^{-1}}{\sigma^2} - \frac{P_t(\phi\phi^T)^{-1}\phi D_i D_i^T \phi^T (\phi\phi^T)^{-1}}{\sigma^4 + \sigma^2 P_t D_i^T \phi^T (\phi\phi^T)^{-1} \phi D_i}. \tag{6.32}$$

Also,

$$(\tilde{\mu}_1^i - \tilde{\mu}_0^i) = \phi s - P_b \phi D_i, \tag{6.33}$$

where $P_b = (P_1^0 - P_2^0) + (P_2^1 - P_1^1)$. Using (6.32), (6.33) and the fact that $D_i = \gamma s$, the deflection coefficient $D(\tilde{y}_i)$ can be calculated to be

$$D(\tilde{y}_i) = (1 - P_b\gamma)^2 \frac{\|\hat{P}s\|_2^2}{\sigma^2} - P_t\gamma^2(1 - P_b\gamma)^2 \frac{\|\hat{P}s\|_2^4}{\sigma^2 r_b}, \tag{6.34}$$

where $r_b = \sigma^2 + P_t\gamma^2\|\hat{P}s\|_2^2$.

Proposition 6.1 ([9]) *Suppose that* $\sqrt{\frac{P}{M}}\hat{P}$ *provides an* ε*-stable embedding of* $(\mathscr{S}, \{0\})$. *Then the deflection coefficient at the FC can be approximated as*

$$D_{FC} \approx \alpha \frac{(1 - P_b\gamma)^2}{\gamma^2 P_t + \frac{P}{M}\frac{\sigma^2}{\|s\|_2^2}} + (1 - \alpha)\frac{M}{P}\frac{\|s\|_2^2}{\sigma^2}, \tag{6.35}$$

where $P_b = (P_1^0 - P_2^0) + (P_2^1 - P_1^1)$ *and* $P_t = P_1^0 + P_2^0 + (P_1^0 - P_2^0)^2$.

Proof Using the fact that $\sqrt{\frac{P}{M}}\hat{P}$ provides an ε-stable embedding of $(S, \{0\})$, $D(y_i)$ and $D(\tilde{y}_i)$ can be approximated as

$$D(y_i) = \frac{M}{P}\frac{\|s\|_2^2}{\sigma^2} \tag{6.36}$$

$$D(\tilde{y}_i) = \frac{M}{P}\frac{\|s\|_2^2}{\sigma^2}(1 - P_b\gamma)^2\left(1 - \frac{M}{P}\frac{\|s\|_2^2}{r_b}\gamma^2 P_t\right), \tag{6.37}$$

where $r_b = \sigma^2 + P_t\gamma^2\frac{M}{P}\|s\|_2^2$. Plugging in the above values in $D_{FC} = \alpha D(\tilde{y}_i) + (1 - \alpha)D(y_i)$ yields the desired result. □

Next, let us obtain the deflection coefficient at the eavesdropper.

A Closed-Form Expression of the Deflection Coefficients at the Eavesdropper

As stated earlier, the deflection coefficient of the eavesdropper is

$$D_{EV} = D(\hat{y}_i). \tag{6.38}$$

Next, to calculate $D(\hat{y}_i)$, observe that $\hat{y}_i$ is distributed as a multivariate Gaussian mixture with

$$\hat{\mu}_0^i = \alpha(P_1^0 - P_2^0)\phi D_i \tag{6.39}$$

$$\hat{\mu}_1^i = \alpha(P_1^1 - P_2^1)\phi D_i + \phi s \tag{6.40}$$

$$\hat{\Sigma}_0^i = \sigma^2\phi\phi^T + \sum_{j=1}^{3} p_j^0(\hat{\mu}_0^i(j) - \hat{\mu}_0^i)(\hat{\mu}_0^i(j) - \hat{\mu}_0^i)^T, \tag{6.41}$$

with $\hat{\mu}_0^i(1) = \phi D_i$, $\hat{\mu}_0^i(2) = -\phi D_i$, $\hat{\mu}_0^i(3) = 0$, $p_1^0 = \alpha P_1^0$, $p_2^0 = \alpha P_2^0$ and $p_3^0 = 1 - \alpha(P_1^0 - P_2^0)$. Using these values, we state our result in the next proposition.

Proposition 6.2 ([9]) *Suppose that* $\sqrt{\frac{P}{M}}\hat{P}$ *provides an* ε*-stable embedding of* $(\mathscr{S}, \{0\})$*. Then, the deflection coefficient at the eavesdropper can be approximated as*

$$D_{EV} \approx \frac{(1-\alpha P_b \gamma)^2}{\alpha\gamma^2 P_t + \frac{P}{M}\frac{\sigma^2}{\|s\|_2^2}} \tag{6.42}$$

where $P_b = (P_1^0 - P_2^0) + (P_2^1 - P_1^1)$ *and* $P_t = P_1^0 + P_2^0 + (P_1^0 - P_2^0)^2$.

Proof The proof is similar to that of Proposition 6.1. □

In general, there is a trade-off between the detection performance and the security performance of the system. For this purpose, the authors in [9] further explored the problem of system parameter design in a holistic manner.

6.1.3 Optimal System Parameter Design Under Physical Layer Secrecy Constraint

Does Compression Help?

First consider the case where $\alpha P_b \gamma \neq 1$. In this regime, for fixed values of α, P_b, and γ, the deflection coefficient, both at the FC and the eavesdropper, is a monotonically increasing function of the compression ratio. In other words, $\frac{dD_{FC}}{dc} > 0$ and $\frac{dD_{EV}}{dc} > 0$. This suggests that compression improves the security performance at the expense of detection performance. More specifically, the FC would decrease the compression ratio until the physical layer secrecy constraint is satisfied. As a consequence, it will result in performance loss at the FC due to compression. In other words, there is a trade-off between the detection performance and the security performance of the system.

To gain insight into this trade-off, Fig. 6.3 shows the deflection coefficient, both at the FC and at the eavesdropper, as a function of compression ratio (c) and data falsification strength (γ) when $\alpha = 0.3$, $P_1^0 = P_2^1 = 0.8$, $P_2^0 = P_1^1 = 0.1$, and $\frac{\|s\|_2^2}{\sigma^2} = 3$. Next, Fig. 6.4 shows the deflection coefficient, both at the FC and at the eavesdropper, as a function of the fraction of data falsifying nodes (α) and data falsification strength (γ) when $P_1^0 = P_2^1 = 0.8$, $P_2^0 = P_1^1 = 0.1$ and $\frac{M}{P}\frac{\|s\|_2^2}{\sigma^2} = 3$.

It can be seen from Figs. 6.3 and 6.4 that D_{FC} and D_{EV} do not have a nice structure (monotonicity or convexity) with respect to the system parameters and, therefore, it is not an easy task to design the system parameters under an arbitrary physical layer secrecy constraint. This motivated [9] to focus their attention on a particular case, namely on the perfect secrecy regime, i.e., $D_{EV} = 0$. Observe that, $D_{EV} = 0$ if and only if $\alpha P_b \gamma = 1$ (ignoring the extreme conditions such as $c = 0$ or $\gamma = \infty$). Thus, in the perfect secrecy regime D_{EV} is independent of the compression ratio c and the network designer can fix $c = c_{max}$, where the value of c_{max} may be dependent on the

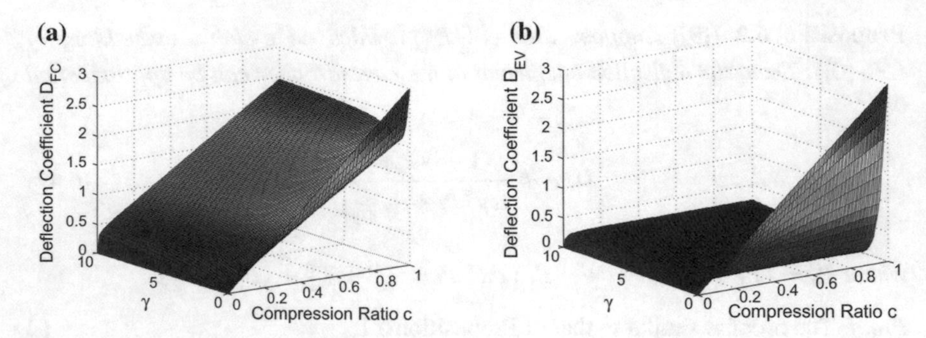

Fig. 6.3 Deflection coefficient analysis. **a** D_{FC} with varying c and γ. **b** D_{EV} with varying c and γ [9]

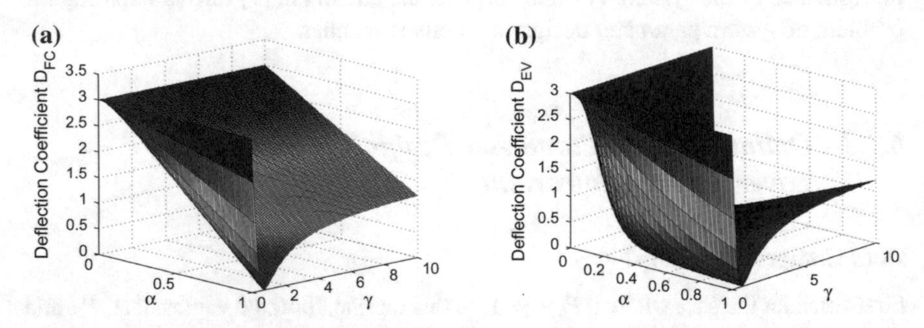

Fig. 6.4 Deflection coefficient analysis. **a** D_{FC} with varying α and γ. **b** D_{EV} with varying α and γ [9]

application of interest. Now, the goal of the designer is to maximize the detection performance D_{FC}, while ensuring perfect secrecy at the eavesdropper.

System Parameter Design in the Perfect Secrecy Regime

The system parameter design problem under the perfect secrecy constraint can be formally stated as

$$\arg\max_{\alpha} D_{FC}(c_{max}, \gamma = 1/(P_b\alpha)). \tag{P6.2}$$

This reduction of the search space, which arises as a natural consequence of the perfect secrecy constraint, has the additional benefit of simplifying the mathematical analysis. To solve (P6.2), let us analyze $D_{FC}(c_{max}, \gamma = 1/(P_b\alpha))$ as a function of α.

Proposition 6.3 ([9]) *In the high signal to noise ratio regime (defined as $\frac{\|s\|_2^2}{\sigma^2} > \frac{P_b^2}{P_t}$), the deflection coefficient at the FC, D_{FC}, is a monotonically decreasing function of the fraction of data falsifying nodes (α) under the perfect secrecy constraint.*

Notice that, Proposition 6.3 suggests that to maximize the deflection coefficient D_{FC} under the perfect secrecy constraint (P6.2), the network designer should choose the value of α as low as possible under the constraint that $\alpha > 0$ and accordingly increase γ to satisfy $\alpha P_b \gamma = 1$. In practice, α_{min} may be dependent on the application of interest.

6.2 Classification via Crowdsourcing with Unreliable Workers

In the recently emerging paradigm of crowdsourcing, crowd workers perform tasks in a distributed manner. These workers often find microtasks tedious and due to lack of motivation fail to generate high-quality work [21]. Therefore, it is important to design crowdsourcing systems with sufficient incentives for workers [11]. The typical incentive for workers is monetary reward, but in [31], it has been found that intrinsic factors such as the challenge associated with the task were a stronger motivation for crowd workers than extrinsic factors such as rewards. Unfortunately, it has been reported that increasing financial incentives increase the number of tasks which the workers take part in but not the per task quality [15]. Recent research, however, suggests that making workers' rewards codependent on each other can significantly increase the quality of their work. This suggests the potency of a *peer-dependent reward scheme* for quality control [6]. In a *teamwork-based scheme*, paired workers are rewarded based on their average work, whereas in a *competition-based* scheme, the paired worker who performs the best gets the entire reward.

Another interesting phenomenon in crowdsourcing is *dependence* of observations among crowd workers [20]. Crowds may share common sources of information, leading to dependent observations among the crowd workers performing the task. Common sources of information have oft-been implicated in the publishing and spread of false information across the Internet, e.g., the premature Steve Jobs obituary, the second bankruptcy of United Airlines, and the creation of black holes by operating the Large Hadron Collider [3]. A graphical model may be used to characterize such dependence among crowd workers [20].

In [32], the coding theory-based scheme used for WSNs in [38] was used for crowdsourcing systems performing inference tasks, namely classification. This approach had twofold benefits [32]. The coding-based scheme helped in designing easy-to-answer binary questions for the humans that improves the performance of individual agents. The second benefit is on the decoding side where the error-correcting code can tolerate a few errors from these agents and, therefore, can make reliable inferences. The remainder of the section is organized as follows: In Sect. 6.2.1, a mathematical model of the crowdsourcing problem is presented and the coding-based approach is described. In Sect. 6.2.2, the crowdsourcing system with independent workers is considered and misclassification performance expressions for both coding- and majority-based approaches are presented. Numerical simulations and experimental results using real data from Amazon Mechanical Turk are also provided. Extensions to peer-dependent reward scheme and dependent crowd workers are briefly discussed.

6.2.1 Coding for Crowdsourcing

In this section, the basic concept of using error-correcting codes to achieve reliable classification in a crowdsourcing system is discussed.

First, let us understand how the DCFECC approach discussed for WSNs in [38] can be used in crowdsourcing systems to design the questions to be posed to the crowd workers. As an example, consider an image to be classified into one of M fine-grained categories. Since object classification is often difficult for machine vision algorithms, human workers may be used for this task. In a typical crowdsourcing microtask platform, a task manager creates simple tasks for the workers to complete, and the results are combined to produce the final result. Due to the low pay of workers and the difficulty of tasks, individual results may be unreliable. Furthermore, workers may not be qualified to make fine-grained M-ary distinctions, but rather can only answer easier questions. Therefore, in the proposed approach, codes are used to design microtasks and decoding is performed to aggregate responses reliably.

Consider the task of classifying a dog image into one of four breeds: Pekingese, Mastiff, Maltese, or Saluki. Since workers may not be canine experts, they may not be able to directly classify and so we should ask simpler questions. For example, the binary question of whether a dog has a snub nose or a long nose differentiates between {Pekingese, Mastiff} and {Maltese, Saluki}, whereas the binary question of whether the dog is small or large differentiates between {Pekingese, Maltese} and {Mastiff, Saluki}. Using a code matrix, we now show how to design binary questions for crowd workers that allow the task manager to reliably infer correct classification even with unreliable workers.

For modeling this system, assume that worker j decides the true class (local decision y_j) with probability p_j and makes the wrong classification with uniform probability:

$$p(y_j|H_m) = \begin{cases} p_j & \text{if } y_j = m \\ \frac{1-p_j}{M-1} & \text{otherwise,} \end{cases} \tag{6.43}$$

For every worker j, let a_j be the corresponding column of $\mathbf{A}$ and recall hypothesis $H_l \in \{H_0, H_1, \cdots, H_{M-1}\}$ is associated with row l in $\mathbf{A}$. The local workers send a binary answer u_j based on decision y_j and column a_j. An illustrative example is shown in Fig. 6.5 for the dog breed classification task above. Let the columns corresponding to the ith and jth workers be $a_i = [1010]'$ and $a_j = [1100]'$, respectively. The ith worker is asked: 'Is the dog small or large?' since the worker is to differentiate between the first (Pekingese) or third (Maltese) breed and the others. The jth worker is asked: 'Does the dog have a snub nose or a long nose?' since the

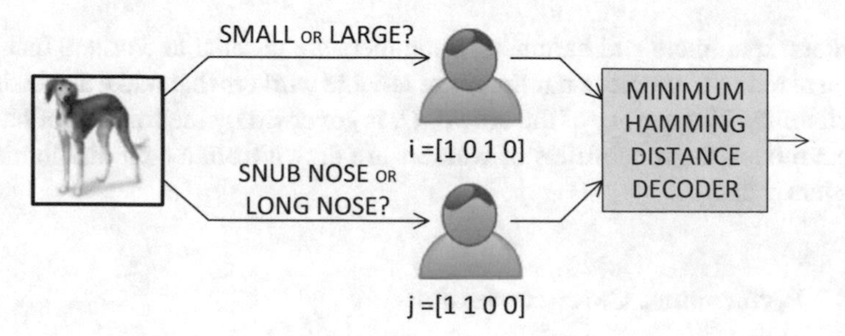

Fig. 6.5 A schematic diagram showing binary questions posed to workers and the decoding rule used by the task manager [32]

worker is to differentiate between the first two breeds (Pekingese, Mastiff) and the others. These questions can be designed using taxonomy and dichotomous keys [25]. Knowing that Pekingese and Maltese are small dogs while the other two breeds are large, we can design the appropriate question as 'Is the dog small or large?' for ith worker whose corresponding column is $a_i = [1010]'$. The task manager makes the final classification as the hypothesis corresponding to the code word (row) that is closest in Hamming distance to the received vector of decisions.

Although distributed classification in sensor networks and in crowdsourcing are structurally similar, an important difference is the anonymity of crowds. Since crowd workers are anonymous, one cannot identify the specific reliability of any specific worker as could be done for a sensor. Hence, in [32], each worker j in the crowd was assumed to have an associated reliability p_j, drawn from a common distribution that characterizes the crowd. Herein, three different crowd models that generate crowd reliabilities were considered: individual and independent crowd workers; crowd workers governed by peer-dependent reward schemes; and crowd workers with common sources of information.

6.2.2 *Crowdsourcing System with Individual Crowd Workers*

In this section, results on the basic crowdsourcing system are presented where independent crowd workers perform the task individually and are rewarded based on their decision only.

6.2.2.1 Model

The system consisting of individual and independent workers can be modeled as one where the workers' reliabilities are drawn i.i.d. from a specific distribution. Two crowd reliability models namely a spammer–hammer model and a beta model were considered in [32]. In a spammer–hammer model, the crowd consists of two kinds of workers: spammers and hammers. Spammers are unreliable workers that make a decision at random, whereas hammers are reliable workers that make a decision with high reliability. The quality of the crowd, Q, is governed by the fraction of hammers. In a beta model, the reliabilities of workers are drawn from a beta distribution with parameters α and β.

6.2.2.2 Performance Characterization

Having defined a coding-based approach to reliable crowdsourcing, its performance is determined in terms of average misclassification probability for classification under minimum Hamming distance decoding. Suppose N workers take part in an M-ary classification task. Let $\mathbf{p}$ denote the reliabilities of these workers, such that p_j for $j = 1, \ldots, N$ are i.i.d. random variables with mean μ. Define this to be an (N, M, μ) crowdsourcing system.

Proposition 6.4 ([32]) *Consider an (N, M, μ) crowdsourcing system. The expected misclassification probability using the code matrix $\mathbf{A}$ is:*

$$P_e(\mu) = \frac{1}{M} \sum_{\mathbf{i},l} \prod_{j=1}^{N} \left[\left(\mu a_{lj} + \frac{(1-\mu)}{(M-1)} \sum_{k \neq l} a_{kj} \right) (2i_j - 1) + (1 - i_j) \right] C_{\mathbf{i}}^l, \tag{6.44}$$

where $\mathbf{i} = [i_1, \cdots, i_N] \in \{0, 1\}^N$ is the received code word and $C_{\mathbf{i}}^l$ is the cost associated with a global decision H_l when the received vector is $\mathbf{i}$. This cost is:

$$C_{\mathbf{i}}^l = \begin{cases} 1 - \frac{1}{\varrho} & \text{if } \mathbf{i} \text{ is in decision region of } H_l \\ 1 & \text{otherwise.} \end{cases} \tag{6.45}$$

where ϱ is the number of decision regions[1] $\mathbf{i}$ belongs to; ϱ can be greater than one when there is a tie at the task manager and the tie-breaking rule is to choose one of them randomly.

Proof Let $P_{e,\mathbf{p}}$ denote the misclassification probability given the reliabilities of the N workers. Then, if u_j denotes the bit sent by the worker j and the global decision is made using the Hamming distance criterion:

$$P_{e,\mathbf{p}} = \frac{1}{M} \sum_{\mathbf{i},l} P(\mathbf{u} = \mathbf{i} | H_l) C_{\mathbf{i}}^l. \tag{6.46}$$

Since local decisions are conditionally independent, $P(\mathbf{u} = \mathbf{i}|H_l) = \prod_{j=1}^{N} P(u_j = i_j|H_l)$. Further,

[1]For each H_l, the set of $\mathbf{i}$ for which the decision H_l is taken is called the decision region of H_l.

$$
\begin{aligned}
P(u_j = i_j|H_l) &= i_j P(u_j = 1|H_l) + (1 - i_j)P(u_j = 0|H_l) &(6.47)\\
&= (1 - i_j) + (2i_j - 1)P(u_j = 1|H_l) &(6.48)\\
&= (1 - i_j) + (2i_j - 1)\sum_{k=1}^{M} a_{kj} P(y_j = k|H_l) &(6.49)\\
&= (1 - i_j) + \left(p_j a_{lj} + \frac{(1 - p_j)}{(M - 1)} \sum_{k \neq l} a_{kj} \right)(2i_j - 1) &(6.50)
\end{aligned}
$$

where y_j is the local decision made by worker j. Since reliabilities p_j are i.i.d. with mean μ, the desired result follows. □

6.2.2.3 Majority Voting

A traditional approach in crowdsourcing has been to use a majority vote to combine local decisions; we also derive its performance for purposes of comparison. For M-ary classification, each worker's local decision is modeled as $\log_2 M$-bit valued, but since workers only answer binary questions, the N workers are split into $\log_2 M$ groups with each group sending information regarding a single bit. For example, consider the dog breed classification task described before which has $M = 4$ classes. Let us represent the classes by 2-bit numbers as follows: Pekingese is represented as '00', Mastiff as '01', Maltese as '10', and Saluki as '11'. The N crowd workers are split into 2 groups. In traditional majority vote, since the workers are asked M-ary questions, each worker first identifies his/her answer. After identifying his/her class, the first group members send the first bit corresponding to their decisions, while the second group members send the second bit of their decisions. The task manager uses a majority rule to decide each of the $\log_2 M$ bits separately and concatenates to make the final classification. Suppose N is divisible by $\log_2 M$.

Proposition 6.5 ([32]) *Consider an (N, M, μ) crowdsourcing system. The expected misclassification probability using majority rule is:*

$$
P_e(\mu) = 1 - \frac{1}{M}\left[1 + S_{\tilde{N},(1-q)}\left(\frac{\tilde{N}}{2}\right) - S_{\tilde{N},q}\left(\frac{\tilde{N}}{2}\right)\right]^{\log_2 M}, \qquad (6.51)
$$

where $\tilde{N} = \frac{N}{\log_2 M}$, $q = \frac{M(1-\mu)}{2(M-1)}$, and $S_{N,p}(\cdot)$ is the survival function (complementary cumulative distribution function) of the binomial random variable $\mathscr{B}(N, p)$.

Proof In a majority-based approach, $\tilde{N} = \frac{N}{\log_2 M}$ workers send information regarding the ith bit of their local decision, $i = 1, \ldots, \log_2 M$. For a correct global decision, all bits have to be correct. Consider the ith bit and let $P^i_{c,\mathbf{p}}$ be the probability of the ith bit being correct given the reliabilities of the $\tilde{N}$ workers sending this bit. Then,

$$P_{c,\mathbf{p}}^{i} = \frac{P_d + 1 - P_f}{2}, \tag{6.52}$$

where P_d is the probability of detecting the ith bit as '1' when the true bit is '1' and P_f is the probability of detecting the ith bit as '1' when the true bit is '0'. Note that '0' and '1' are equiprobable since all hypotheses are equiprobable. Under majority rule for this ith bit,

$$P_d = \sum_{j=\lfloor \frac{\tilde{N}}{2}+1 \rfloor}^{\tilde{N}} \sum_{\forall G_j} \prod_{k \in G_j} \left(1 - \frac{M(1-p_k)}{2(M-1)}\right) \prod_{k \notin G_j} \frac{M(1-p_k)}{2(M-1)}, \tag{6.53}$$

where G_j is a set of j out of $\tilde{N}$ workers who send bit value '1' and $\frac{M(1-p_k)}{2(M-1)}$ is the probability of the kth worker making a wrong decision for the ith bit. Similarly,

$$P_f = \sum_{j=\lfloor \frac{\tilde{N}}{2}+1 \rfloor}^{\tilde{N}} \sum_{\forall G_j} \prod_{k \in G_j} \frac{M(1-p_k)}{2(M-1)} \prod_{k \notin G_j} \left(1 - \frac{M(1-p_k)}{2(M-1)}\right). \tag{6.54}$$

Now the overall probability of correct decision is given by $P_{c,\mathbf{p}} = \prod_{i=1}^{\log_2 M} P_{c,\mathbf{p}}^{i}$. Since reliabilities are i.i.d., the expected probability of correct decision P_c is:

$$P_c = \prod_{i=1}^{\log_2 M} E[P_{c,\mathbf{p}}^{i}], \tag{6.55}$$

where expectation is with respect to $\mathbf{p}$. Since reliabilities are i.i.d.:

$$E[P_d] = \sum_{j=\lfloor \frac{\tilde{N}}{2}+1 \rfloor}^{\tilde{N}} \binom{\tilde{N}}{j} (1-q)^j q^{(\tilde{N}-j)} = S_{\tilde{N},(1-q)}\left(\frac{\tilde{N}}{2}\right), \tag{6.56}$$

$$E[P_f] = \sum_{j=\lfloor \frac{\tilde{N}}{2}+1 \rfloor}^{\tilde{N}} \binom{\tilde{N}}{j} q^j (1-q)^{(\tilde{N}-j)} = S_{\tilde{N},q}\left(\frac{\tilde{N}}{2}\right). \tag{6.57}$$

Using (6.52), (6.55), (6.56), and (6.57), we get the desired result. □

6.2.2.4 Performance Evaluation

In [32], an ordering principle for the quality of crowds was defined in terms of the quality of their distributed inference performance. This is a valuable concept since it provides us a tool to evaluate a given crowd. Such a valuation could be used by the task manager to pick the appropriate crowd for the task based on the performance requirements. For example, if the task manager is interested in constraining the misclassification probability of his/her task to ε while simultaneously minimizing the required crowd size, the above expressions can be used to choose the appropriate crowd.

Theorem 6.2 ([32], Ordering of Crowds) *Consider crowdsourcing systems involving crowd $\mathscr{C}(\mu)$ of workers with i.i.d. reliabilities with mean μ. Crowd $\mathscr{C}(\mu)$ performs better than crowd $\mathscr{C}(\mu')$ for classification if and only if $\mu > \mu'$.*

Proof As can be observed from Propositions 6.4 and 6.5, the average misclassification probabilities depend only on the mean of the reliabilities of the crowd. Therefore, it follows that crowd $\mathscr{C}(\mu)$ of workers with i.i.d. reliabilities with mean μ performs better for classification than crowd $\mathscr{C}(\mu')$ of workers with i.i.d. reliabilities with mean μ' as $\mu > \mu'$. □

Since the performance criterion is average misclassification probability, this can be regarded as a weak criterion of crowd-ordering in the mean sense. Thus, with this crowd-ordering, better crowds yield better performance in terms of average misclassification probability. Indeed, misclassification probability decreases with better quality crowds.

Consider a simulated crowdsourcing system as follows: $N = 10$ workers take part in a classification task with $M = 4$ equiprobable classes. A good code matrix $\mathbf{A}$ is found by simulated annealing [38]:

$$\mathbf{A} = [5, 12, 3, 10, 12, 9, 9, 10, 9, 12]. \tag{6.58}$$

Here, code matrices are represented as a vector of M bit integers. Each integer r_j represents a column of the code matrix $\mathbf{A}$ and can be expressed as $r_j = \sum_{l=0}^{M-1} a_{lj} \times 2^l$. For example, the integer 5 in column 1 of $\mathbf{A}$ represents $a_{01} = 1, a_{11} = 0, a_{21} = 1$ and $a_{31} = 0$.

Consider the setting where all the workers have the same reliability $p_j = p$. Figure 6.6 shows the probability of misclassification as a function of p. As is apparent, the probability of misclassification reduces with reliability and approaches 0 as $p \to 1$, as expected.

Now the performance of the coding-based approach is compared to the majority-based approach. Figure 6.7 shows the misclassification probability as a function of crowd quality for $N = 10$ workers taking part in an $(M = 4)$-ary classification task. The spammer–hammer model, where spammers have reliability $p = 1/M$ and hammers have reliability $p = 1$, is used. The figure shows a slight improvement in performance over the majority vote when code matrix (6.58) is used.

Now consider a larger system with increased M and N. A good code matrix $\mathbf{A}$ for $N = 15$ and $M = 8$ is found by cyclic column replacement:

$$\mathbf{A} = [150, 150, 90, 240, 240, 153, 102, 204, 204, 204, 170, 170, 170, 170, 170]. \tag{6.59}$$

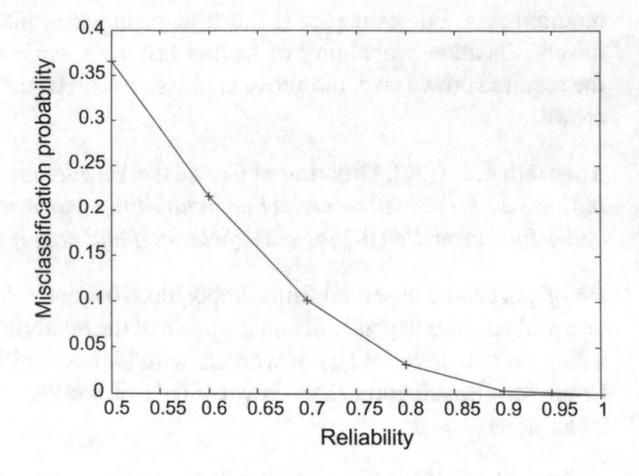

Fig. 6.6 Coding-based crowdsourcing system misclassification probability as a function of worker reliability [32]

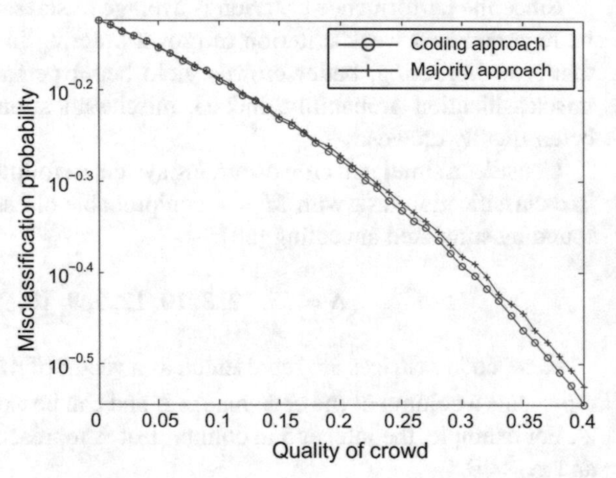

Fig. 6.7 Misclassification probability as a function of crowd quality using coding- and majority-based approaches with the spammer–hammer model, ($M = 4, N = 10$) [32]

The code matrix for the system with $N = 90$ and $M = 8$ was formed suboptimally by concatenating the columns of (6.59) six times. Due to the large system size, it is computationally very expensive to optimize for the code matrix using either the simulated annealing or cyclic column replacement methods. Therefore, in [32], the columns of (6.59) were concatenated. This can be interpreted as a crowdsourcing system of 90 crowd workers consisting of 6 sub-systems with 15 workers each which are given the same task and their data is fused together. In the extreme case, if each of these sub-systems was of size one, it would correspond to a majority vote where all the workers are posed the same question. Figure 6.8 shows the performance when $M = 8$ and N takes the two values: $N = 15$ and $N = 90$. These figures suggest that the gap in performance generally increases for larger system size. Similar observations hold for the beta model of crowds; see Figs. 6.9 and 6.10. It was concluded in [32] that good codes perform better than majority vote as they diversify the binary questions which are asked to the workers.

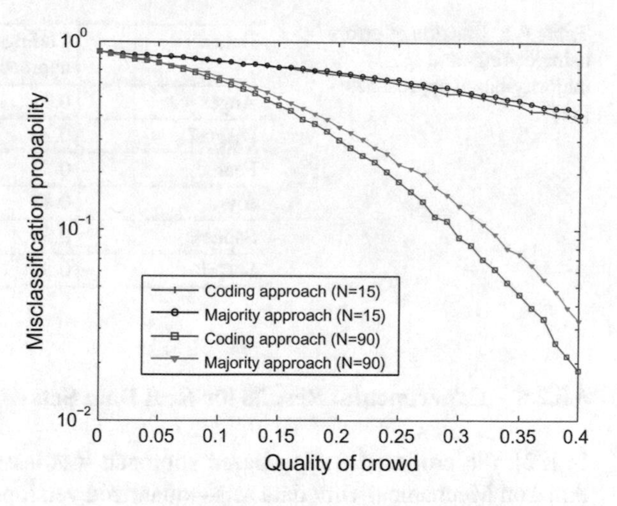

Fig. 6.8 Misclassification probability as a function of crowd quality using coding- and majority-based approaches with the spammer–hammer model, $(M = 8)$ [32]

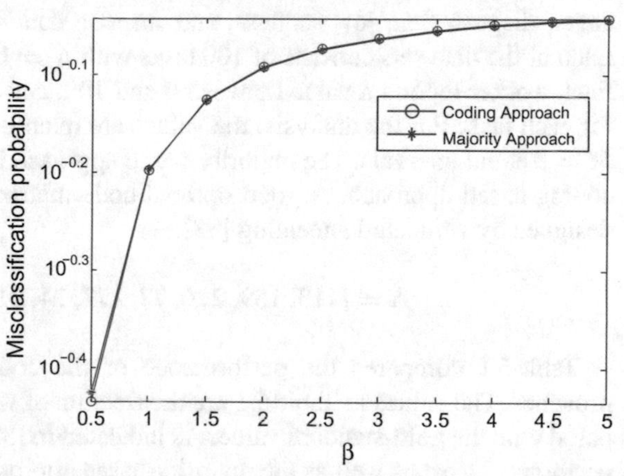

Fig. 6.9 Misclassification probability as a function of β using coding- and majority-based approaches with the Beta$(\alpha = 0.5, \beta)$ model, $(M = 4, N = 10)$ [32]

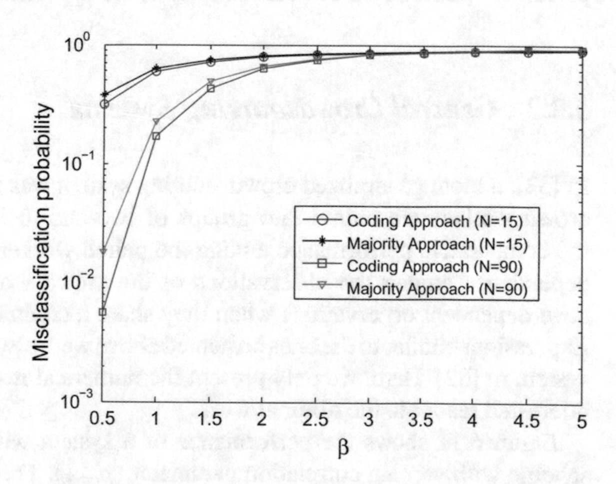

Fig. 6.10 Misclassification probability as a function of β using coding- and majority-based approaches with the Beta$(\alpha = 0.5, \beta)$ model, $(M = 8)$ [32]

Table 6.1 Fraction of errors using coding- and majority-based approaches [32]

Data set	Coding-based approach	Majority-based approach
Anger	0.31	0.31
Disgust	0.26	0.20
Fear	0.32	0.30
Joy	0.45	0.47
Sadness	0.37	0.39
Surprise	0.59	0.63

6.2.2.5 Experimental Results for Real Data Sets

In [32], the proposed coding-based approach was tested on six publicly available Amazon Mechanical Turk data sets—quantized versions of the data sets in [30]: the anger, disgust, fear, joy, sadness, and surprise data sets of the affective text task. Each of the data sets consists of 100 tasks with $N = 10$ workers taking part in each. Each worker reports a value between 0 and 100, and there is a gold-standard value for each task. For the analysis, the values are quantized by dividing the range into $M = 8$ equal intervals. The majority-based approach is compared with the proposed coding-based approach. A good optimal code matrix for $N = 10$ and $M = 8$ was designed by simulated annealing [38]:

$$\mathbf{A} = [113, 139, 226, 77, 172, 74, 216, 30, 122]. \tag{6.60}$$

Table 6.1 compares the performance of the coding- and majority-based approaches. The values in Table 6.1 are the fraction of wrong decisions made, as compared with the gold-standard value. As indicated in [32], the coding-based approach performs at least as well as the majority-based approach in 4 of 6 cases considered.

6.2.3 General Crowdsourcing Systems

In [32], a more generalized crowdsourcing system was also considered wherein the crowd workers are paired into groups of two and their reward value is based on the comparative performance among the paired workers [6]. They also considered dependence among the observations of the crowd workers. Crowd workers may have dependent observations when they share a common information source [20]. Expressions similar to the ones presented above were also derived for this generalized system in [32]. Here, we only present the numerical results from [32] and refer the interested reader to the original work.

Figure 6.11 shows the performance of a system with a peer-dependent reward scheme with varying correlation parameter (ρ_{corr}). The plots are for a system with

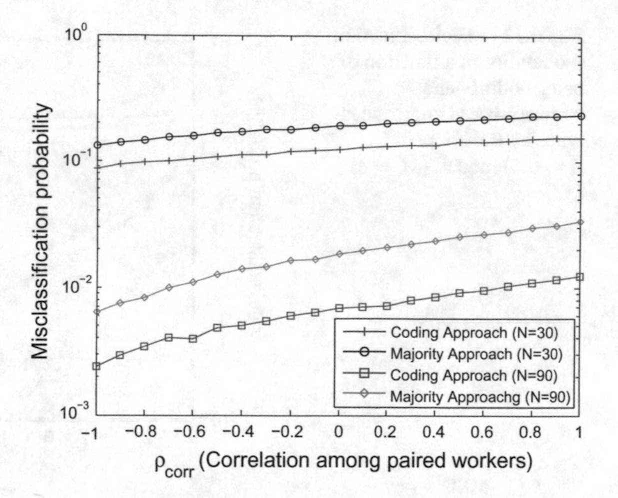

Fig. 6.11 Misclassification probability as a function of ρ_{corr} using coding- and majority-based approaches with the Beta($\alpha = 0.5$, $\beta = 0.5$) model, ($M = 8$) [32]

N crowd workers and $M = 8$, i.e., performing 8-ary classification. As the figure suggests, the performance of the system improves when the workers are paired in the right manner. Note that when the crowdsourcing system with peer-dependent reward scheme has $\rho_{corr} = 0$, it reduces to the system with individual crowd workers considered in Sect. 6.2.2. Figure 6.11 includes the system with independent crowd workers as a special case when $\rho_{corr} = 0$. Also, these results suggest that having a negative correlation among workers results in better performance while a positive correlation deteriorates system performance. This observation can be attributed to the amount of diversity among the crowd workers. When $\rho_{corr} = 1$, the workers are correlated with each other while a correlation value of $\rho_{corr} = -1$ corresponds to a diverse set of workers which results in improved performance.

Figure 6.12 shows the performance of the system as a function of the concentration parameter (κ), which characterizes the amount of dependency in observations: a high value of κ implies independent observations while $\kappa = 0$ implies completely dependent observations (all workers have a single source of information). The plots in Fig. 6.12 are for a system with uncorrelated workers (no peer-dependent reward scheme) performing an 8-ary classification ($M = 8$).

As expected, one can observe that the system performance improves when the observations become independent. Also, the performance gap between majority- and coding-based approaches increases as the observations become independent. When the observations are dependent, all workers have similar observations on an average, so posing similar questions (majority-based approach) will perform similarly to posing diverse questions (coding-based approach). On the other hand, when the observations are independent, it is more informative to pose diverse questions to these workers as done in the coding-based approach. Therefore, it was inferred in [32] that *diversity is good* and the benefit of using coding increases when the observations are more diverse.

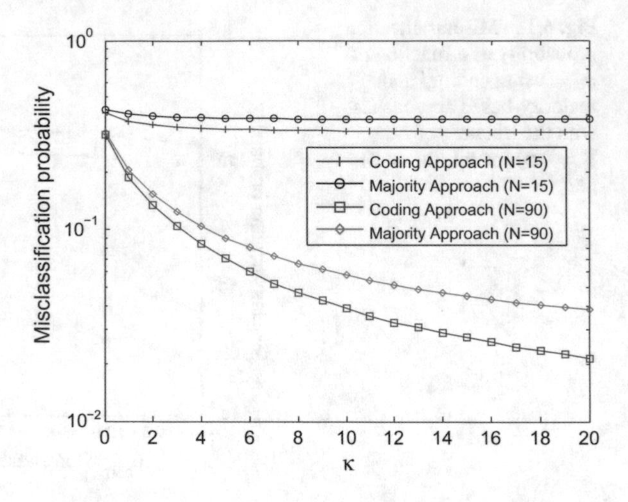

Fig. 6.12 Misclassification probability as a function of κ using coding- and majority-based approaches with the Beta($\alpha = 0.5$, $\beta = 0.5$) model, ($M = 8$) [32]

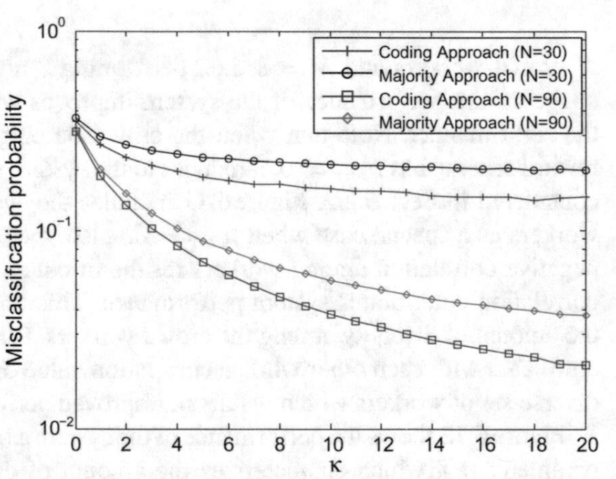

Fig. 6.13 Misclassification probability as a function of κ using coding- and majority-based approaches with the Beta($\alpha = 0.5$, $\beta = 0.5$) model and $\rho = -0.5$, ($M = 8$) [32]

Similar observations were also made for a system consisting of peer-dependent reward scheme with dependent observations. Figure 6.13 shows the performance of such a system using both majority- and coding-based approaches. The plots are for a system with N crowd workers performing an 8-ary classification task ($M = 8$). The group reliability distribution is assumed to be beta distribution and the correlation parameter is $\rho_{corr} = -0.5$.

6.3 Human–Machine Inference Networks

Another direction of recent research that is relevant to the theme discussed in this book is the paradigm of Human–Machine Inference Networks (HuMaINs). In traditional economics, cognitive psychology, and artificial intelligence (AI) literature, the problem-solving or inference process is described in terms of searching a problem space, which consists of various states of the problem, starting with the initial state and ending at the goal state which one would like to reach [1]. Each path from the initial state represents a possible strategy which can be used. These paths could either lead to the desired goal state or to other non-goal states. The paths from the initial state that lead to the goal state are called the solution paths. There could be multiple such paths between the initial and the goal state which are all solutions to the problem. In other words, there are multiple ways to solve a given problem. The problem-solving process is to identify the optimal (under a given constraint) solution path among the multiple solution paths emanating from the initial state and reaching the goal state.

The first step for such a search is to determine the set of available strategies, i.e., the strategy space. The second step is to evaluate the strategies to determine the best strategy as the solution. In traditional economic theory, a rational decision maker is assumed to have the knowledge of the set of possible alternatives,[2] has the capability to evaluate the consequences of each alternative, and has a utility function which he/she tries to maximize to determine the optimal strategy [14]. However, it is widely accepted that humans are not rational but are bounded rational agents. Under the bounded rationality framework [14, 28], decision makers are cognitively limited and have limited time, limited information, and limited resources. The set of alternatives is not completely known *a priori* nor are the decision makers perfectly aware of the consequences of choosing a particular alternative. Therefore, the decision maker might not always determine the best strategy for solving the problem.

On the other hand, machines/sensors are *rational* in the sense that they have stronger/larger memory for storing alternatives and have the computational capability to more accurately evaluate the consequences of a particular alternative. Therefore, a machine can aid a human in fast and accurate problem-solving. This leads us to this framework for human–machine collaboration for problem-solving. In this section, we discuss this collaboration framework presented in [35] for inference by defining the Human–Machine Inference Networks (HuMaINs) and discussing the research challenges associated with developing such a framework. The three basic threads of research in this area are defined.

6.3.1 *HuMaIN Framework*

Figure 6.14 presents a typical Human–Machine Inference Network (HuMaIN). A typical HuMaIN consists of a social network where humans exchange subjective opinions among themselves, and a machine network where machines exchange objective measurements among them. Moreover, due to the interaction between social and machine networks, the behavioral characteristics of humans determine algorithms adopted by machines and these algorithms in turn affect the behavior of

[2]The terms strategy and alternative are used interchangeably.

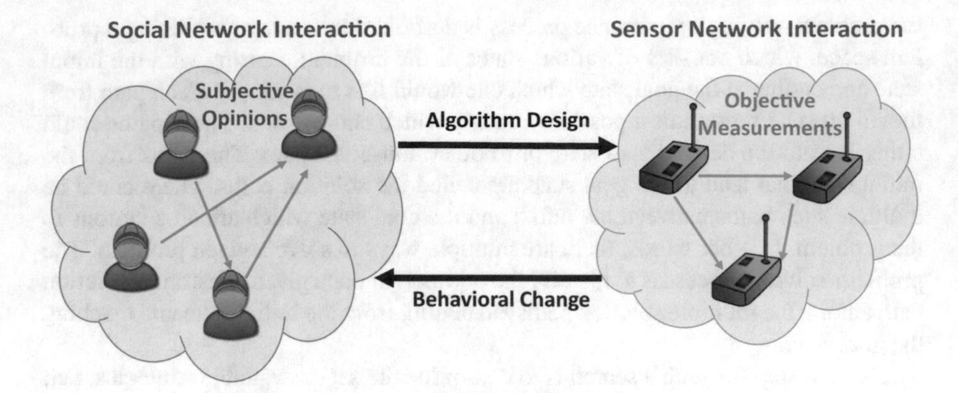

Fig. 6.14 Notional HuMaINs architecture [35]

humans. Therefore, an intelligent collaboration of humans and machines can deliver improved results, by exploiting the strengths of humans and machines.

There are three major directions of research that fall under the HuMaIN paradigm: (1) architecture, (2) algorithms, and (3) applications.

6.3.2 Architecture

Several control architectures involve the interaction of an autonomous system with one or more human agents. Examples of such an architecture include fly-by-wire aircraft control systems (interacting with a pilot), automobiles with driver assistance systems (interacting with a driver), and medical devices (interacting with a doctor, nurse, or patient) [13]. The success of such architectures depends not only on the autonomous system, but also on the actions of the human agents. The goal is to develop a human decision-making framework that quantifies the human representation in the decision-making task under uncertainty and also develop an estimator for the model parameters. The framework should also provide a common ontology for humans and machines to share relevant information about the task. By estimating the parameters, a machine can access this representation and potentially improve its performance. In control systems terminology, the model and associated estimator should form a plant–observer pair for human decision making that can be used for system design [24]. Incorporating these ideas into the feedback control framework will require new results and theory to provide performance guarantees.

The authors in [35] classify the architectures into three categories: (1) architectures where humans directly control the autonomous system, (2) architectures where the autonomous system monitors humans and takes actions if required, and (3) a combination of 1 and 2. In order to achieve the goal of HuMaINs, it is critical to build an architecture that lends itself to a blend of human and machine decision making. In [19], it is stated that to create a state-of-the-art operator environment for modern automation systems, continued technology development is needed in three major areas: decision support tools; ergonomics and visualization technologies; and ease-of-use of complex systems. Research focusing on building such systems falls under the research paradigm of human-in-the-loop cyber-physical system (HiLCPS) [26]. As Schirner et al. [26] state, designing and implementing a HiLCPS poses challenges that require multi-disciplinary research to solve these challenges. Research in the areas of control systems, human–computer interface (HCI), and systems design, together will drive the design of a HuMaIN architecture.

6.3.3 Algorithms

The key research area for HuMaINs is the development of new algorithms that deal with the human-behavioral data while also handling the presence of sensors that can be prone to errors/attacks. Research that deals with the human-behavioral data falls under the paradigm of an emerging research area called *behavioral signal processing* [16]. Behavioral signal processing (BSP) deals with human-behavioral signals. It is defined as processing of human action and behavior data for meaningful analysis to ensure timely decision making and intervention (action) by collaborative integration of human expertise with automated processing. The goal is to support and not supplant humans [16]. The core elements include quantitative understanding of human behavior and mathematical modeling of interaction dynamics. Narayan and Georgiou describe the elements of BSP by using speech and spoken language communication for measuring and modeling human behavior [16] (Fig. 6.15).

There are two specific research directions discussed in [35] while developing BSP algorithms for HuMaINs:

1. Develop mathematical models of human decision making using statistical modeling techniques, in close collaboration with cognitive psychologists and
2. Design robust fusion algorithms that handle unreliable data from the agents as modeled by the above-developed models.

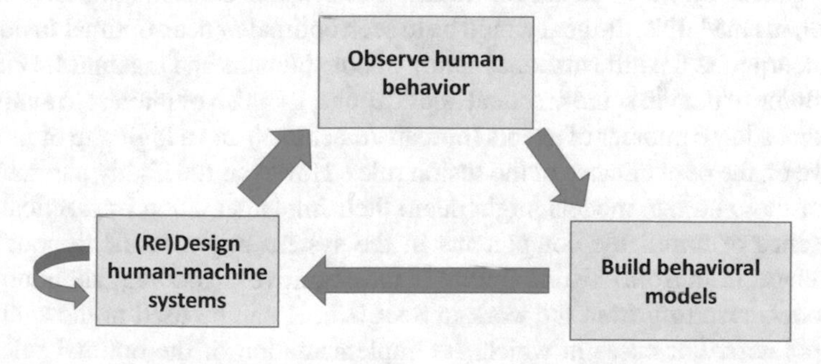

Fig. 6.15 General approach for the design and analysis of HuMaINs [35]

These problems have both theoretical and implementation challenges. Both these research problems are further discussed in some detail below.

6.3.3.1 Statistical Modeling of Human Behavior

The first step toward developing efficient systems containing humans and machines is to develop appropriate models that characterize their behavior. While statistical models exist that characterize the machine observations, researchers have not extensively investigated the modeling of decisions and subjective confidences on multi-hypothesis tasks, or on tasks in which human decision makers can provide imprecise (i.e., vague) decisions. Both of these task types, however, are important in the many applications of HuMaINs. In the preliminary work [36], a comparative study between people and machines for the task of decision fusion has been performed. It was observed that the behavior between people and machines is different since the optimal fusion rule is a deterministic one while people typically use non-deterministic rules which depend on various factors. Based on these observations, a hierarchical Bayesian model was developed to address the observed behavior of humans. This model captured the differences observed in people at individual level, crowd level, and population level. Moving forward, for individual human decision-making models, tools from bounded rationality framework [28] and rational inattention theory [29] can be used in building a theory. Experiments with human subjects can be designed to model the cognitive mechanisms which govern the generation of decisions and decision confidences as they pertain to the formulation of precise and imprecise decisions. One can also build models that consider the effect of stress, anxiety, and fatigue in the cognitive mechanisms of human decision making, decision confidence assessment, and response time (similar to [23, 39]).

6.3.3.2 Design of Robust Algorithms

The next step after deriving probabilistic models of human decision-making is to develop efficient fusion algorithms for collaborative decision making. This has been the major focus of this book but for traditional networks consisting of only sensors. However, in HuMaINs, the goal would be to seek optimal or near-optimal fusion rules which incorporate the informational nature of both humans and machines. Due to the large volume of data in some practical applications, it is also of interest to analyze the effects that a large number of agents (humans/machines) and a high rate of incoming data have on the performance of the fusion rules. However, the highly parameterized nature of these human models might deem their implementation impractical. Also, the presence of unreliable components in the system might result in poor fusion performance. Data from existing studies in the cognitive psychology literature along with models resulting from the work in Sect. 6.3.3.1 can be used in the analysis of these operators. For cases in which the implementation of the optimal rule is not feasible, one must investigate the use of adaptive fusion rules that attempt to learn the parameters of the optimal fusion rule online. Also, for the design of simple and robust algorithms, ideas from coding theory can be used similar to the reliable crowdsourcing results such as in [34].

For the development of future systems consisting of humans and machines, the methodology described above needs to be implemented. First, statistical models of humans should be developed, which are then used to optimize the machines in the system. Due to the presence of potential unreliable agents, one has to also take into consideration the robustness of the systems while developing such large-scale systems. For example, [8, 33, 34] demonstrated the utility of statistical learning techniques and tools from coding theory to achieve reliable performance from unreliable agents.

6.3.4 Applications

Another major driver for the development of large-scale HuMaINs is the application areas. Each application area that deals with human–machine collaboration has its own specific nuances that drive the architecture and algorithmic solutions. In this paper, we discuss four extremely important and timely application areas: education, autonomous vehicles, health care, and science. We discuss their associated research problems in the context of HuMaINs.

6.3.4.1 Education

Human-in-the-loop system can have a significant impact in education domain. The research field of intelligent tutoring systems (ITS) is attempting to design computer systems that can provide immediate and customized instruction or feedback to learners, with intervention from a human teacher. They are enabled to serve as complementing a human teacher and ensure personalized and adaptive learning at scale to every learner. While ITS research has been active for several decades [2, 4], recent advancements in AI and big data research has enabled increasingly more human-like interactions with computers giving rise to interactive, engaging, and immersive tutoring systems. A typical ITS consists of four basic components [17, 18]: Domain model, Student model, Tutoring model, and User interface model. The Domain Model contains the skills, concepts, rules, and/or problem-solving strategies of the domain to be learned. The Student/Learner Model is an overlay on the domain model, and it models the student's cognitive and affective understanding of the domain and their evolution during the learning process. The tutor model represents the tutoring strategies and actions that are dependent on the domain model and the specific learner. The user interface component integrates the other three to ensure interaction with the user and learning advances as planned.

With respect to the HuMaIN paradigm discussed in Sect. 6.3, the domain model represents the task or goal of a HuMaIN, research on the user/learner model represents the human aspect of HuMaINs and modeling human behavior, the tutoring model represents the machine aspect of HuMaINs and designing of robust inference algorithms, and the user interface model represents the architectural research of designing HuMaINs.

6.3.4.2 Autonomous Vehicles

Detection, localization, control, and path planning are essential components of autonomous vehicle design [12, 37]. These tasks focus on sensing and interacting with the physical world through sensors and actuators. Although autonomous vehicles can be a game changer, there are still many obstacles holding back their deployment in practice. Autonomous nature of these systems makes them quite vulnerable to cyber-attacks. A solution is to employ human-in-the-loop systems (semi-autonomous driving) for safe and intelligent autonomous vehicle operation. Such systems would require joint environment-driver state sensing, inference, and shared control and new metrics to characterize safety. The measures of system safety should take into account human performance in response to unexpected hazardous events, and human decision making during vehicle malfunctions caused by cyber-attacks. Furthermore, allowing communication among multiple self-driving cars can enable collective intelligence in such systems, however, would require the design of robust communication protocols.

6.3.4.3 Health Informatics

Automated inference using machine learning (ML) for health care holds enormous potential to increase quality, efficacy and efficiency of treatment and care [22]. Automatic approaches greatly benefit from big data with many training samples. Several tasks in medical domain have high dimensional complex data, where the inclusion of a human is impossible and ML shows impressive results. On the other hand, for certain tasks one is confronted with a small number of data sets or rare events, where ML approaches suffer from insufficient training samples. Furthermore, in health care, decisions made by machines can have serious consequences and necessitate the incorporation of human experts' domain knowledge. There is also a growing trend of litigation requiring the need to bring human in the loop. This makes doctor-in-the-loop systems to be a perfect candidate for health care. Designing such systems would require devising ML approaches that can interact with human agents (doctors) and can optimize their learning behavior through these interactions. Furthermore, unlike current black box-like ML approaches, we need interpretable ML models for health care so that these systems can become transparent to earn experts' trust and be adopted in their workflow.

6.3.4.4 Scientific Discovery

Scientific research spans problems and challenges ranging from screening of novel materials with desired performance in material science, optimizing the analysis of the Higgs boson in high energy physics, tracking of extreme weather phenomena in climate science. Currently, the role of machines in accelerating science has been limited to solving a well-defined task where the data and techniques are given to them by the scientists. This limits our ability to tackle problems where not only the complexity of the data but the questions and the tasks itself challenge our human capabilities to make discoveries [5]. HuMaINs can play an important role in scientific research and become crucial as more interdisciplinary science questions are tackled. This would require advancing machine learning techniques to do independent inquiry, proactive learning, and deliberative reasoning in the presence of hypotheses, domain knowledge, and insights provided by the scientists.

6.4 Other Future Research Directions

Although studying Byzantine attacks and defenses in distributed inference networks has been an active area of research in the past several years, there are several interesting future directions for research. Most of the existing works on distributed inference with Byzantines consider a static scenario where sensors and FC are stationary. Decentralized networks, such as mobile ad hoc networks, have found increasingly wide applications in military communication and operations, search and rescue operations, wireless Internet of things. Several of these application scenarios employ distributed inference techniques to achieve the end goal. Attack and defense issues of mobile ad hoc inference networks are understudied compared to their centralized (and static) counterparts. Furthermore, mobile ad hoc inference networks are expected to have time-varying environments due to, e.g., the mobility of sensors and/or FC. Consequently, efficient attack and defense schemes should dynamically adapt to time-varying scenarios. Finally, ensuring reliability in the presence of Byzantine attacks in such emerging application scenarios will require more sophisticated design across multiple layers of the networking protocol stack: advanced distributed inference at the physical layer, sophisticated network coding schemes for large networks, and a variety of cryptographic techniques for different applications.

Also, moving forward, the social aspect of HuMaINs, with multiple human and machine components interacting, such as in Internet of things (IoT) systems is a direction for future research. Other important research questions are as follows. Can better cognitive and attention models of human crowd workers provide better insight and design principles?

6.5 Summary

In this chapter, we presented results from some recent innovative research paradigms that are similar in problem setup or solution methodology to the problems focused in this book. We showed that Byzantines have been helpful for network in countering secrecy by using them to send falsified data to the eavesdropper. We also presented the successful transfer of the coding approaches typically used for WSNs, for reliable classification in crowdsourcing systems. And finally, we presented an overview of Human–Machine Inference Networks proposed in [35]. A holistic research initiative including researchers from signal processing, security, networking, and psychology, is needed to empower this new field of HuMaIN research.

References

1. Anderson JR (2010) Cognitive psychology and its implications. Worth Publishers, New York
2. Anderson JR, Boyle CF, Reiser BJ (1985) Intelligent tutoring systems. Science 228(4698):456–462
3. Berti-Equille L, Sarma AD, Dong X, Marian A, Srivastava D (2009) Sailing the information ocean with awareness of currents: Discovery and application of source dependence. In: Proceedings of the conference innovations data systems research
4. Corbett AT, Koedinger KR, Anderson JR (1997) Intelligent tutoring systems. In: Helander M, Landauer TK, Prabhu P (eds) Handbook of human-computer interaction. Elsevier Science B. V., pp 849–874
5. Gil Y (2017) Thoughtful artificial intelligence: forging a new partnership for data science and scientific discovery. IOS Press, Preprint, Data Science
6. Huang SW, Fu WT (2013) Don't hide in the crowd! increasing social transparency between peer workers improves crowdsourcing outcomes. In: Proceedings of the SIGCHI conference on human factors in computing systems (CHI 2013), pp 621–630. https://doi.org/10.1145/2470654.2470743
7. Kailkhura B, Nadendla VSS, Varshney PK (2015a) Distributed inference in the presence of eavesdroppers: a survey. IEEE Commun Mag 53(6):40–46
8. Kailkhura B, Vempaty A, Varshney PK (2015b) Distributed inference in tree networks using coding theory. IEEE Trans Signal Process 63(14):3715–3726. https://doi.org/10.1109/TSP.2015.2434326
9. Kailkhura B, Wimalajeewa T, Varshney PK (2017) Collaborative compressive detection with physical layer secrecy constraints. IEEE Trans Signal Process 65(4):1013–1025
10. Kay SM (1998) Fundamentals of statistical signal processing: detection theory. Prentice Hall Signal Processing Series, vol 2. A. V. Oppenheim, Ed. Prentice Hall PTR
11. Law E, von Ahn L (2011) Human computation. Morgan & Claypool Publishers
12. Levinson J, Askeland J, Becker J, Dolson J, Held D, Kammel S, Kolter JZ, Langer D, Pink O, Pratt V, Sokolsky M, Stanek G, Stavens D, Teichman A, Werling M, Thrun S (2011) Towards fully autonomous driving: Systems and algorithms. In: Proceedings of the 2011 IEEE intelligent vehicles symposium (IV), pp 163–168. https://doi.org/10.1109/IVS.2011.5940562
13. Li W, Sadigh D, Sastry SS, Seshia SA (2014) Synthesis for human-in-the-loop control systems. In: Proceedings of the 20th international conference tools and algorithms for the construction and analysis of systems (TACAS), pp 470–484
14. March JG, Simon HA (1958) Organizations. Wiley, New York

15. Mason W, Watts DJ (2009) Financial incentives and the "performance of crowds". In: Proceedings of the ACM SIGKDD Workshop Human Computation (HCOMP'09), pp 77–85. https://doi.org/10.1145/1600150.1600175
16. Narayanan S, Georgiou PG (2013) Behavioral signal processing: deriving human behavioral informatics from speech and language. Proc IEEE 101(5):1203–1233. https://doi.org/10.1109/JPROC.2012.2236291
17. Nkambou R, Mizoguchi R, Bourdeau J (2010) Advances in intelligent tutoring systems. Springer Science & Business Media, Heidelberg
18. Nwana HS (1990) Intelligent tutoring systems: an overview. Artif Intell Rev 4(4):251–277
19. Pretlove J, Skourup C (2007) Human in the loop. ABB Rev 1:6–10
20. Qi GJ, Aggarwal CC, Han J, Huang T (2013) Mining collective intelligence in diverse groups. In: Proceedings of the 22nd International Conferernce World Wide Web (WWW'13), pp 1041–1052
21. Quinn AJ, Bederson BB (2011) Human computation: a survey and taxonomy of a growing field. In: Proceedings of the 2011 annual conference on human factors in computing systems (CHI 2011), pp 1403–1412. https://doi.org/10.1145/1978942.1979148
22. Raghupathi W, Raghupathi V (2014) Big data analytics in healthcare: promise and potential. Health Inf Sci Syst 2(1):1–10. https://doi.org/10.1186/2047-2501-2-3
23. Ratcliff R, van Dongen HPA (2011) Diffusion model for one-choice reaction-time tasks and the cognitive effects of sleep deprivation. Proc Natl Acad Sci USA 108(27):11,285–11,290
24. Reverdy PB (2014) Human-inspired algorithms for search: a framework for human-machine multi-armed bandit problems. PhD thesis, Princeton University
25. Rocker J, Yauch CM, Yenduri S, Perkins LA, Zand F (2007) Paper-based dichotomous key to computer based application for biological identification. J Comput Sci Coll 22(5):30–38
26. Schirner G, Erdogmus D, Chowdhury K, Padir T (2013) The future of human-in-the-loop cyber-physical systems. IEEE Comput 46(1):36–45. https://doi.org/10.1109/MC.2013.31
27. Sherman J, Morrison WJ (1950) Adjustment of an inverse matrix corresponding to a change in one element of a given matrix. Ann Math Stat 21(1):124–127. https://doi.org/10.1214/aoms/1177729893
28. Simon HA (1982) Models of bounded rationality: empirically grounded economic reason. MIT Press, Cambridge
29. Sims CA (2003) Implications of rational inattention. J Monet Econ 50(3):665–690. https://doi.org/10.1016/S0304-3932(03)00029-1
30. Snow R, O'Connor B, Jurafsky D, Ng AY (2008) Cheap and fast—but is it good?: Evaluating non-expert annotations for natural language tasks. In: Proceedings of the conference on empirical methods in natural language processing (EMNLP'08), pp 254–263
31. Varshney LR (2012) Participation in crowd systems. In: Proceedings of the 50th Annual Allerton Conference on Communication Control and Computing, pp 996–1001. https://doi.org/10.1109/Allerton.2012.6483327
32. Vempaty A, Varshney LR, Varshney PK (2013) Reliable classification by unreliable crowds. In: Proceedings of the IEEE international conference on acoustics, speech, and signal processing (ICASSP 2013), pp 5558–5562
33. Vempaty A, Han YS, Varshney PK (2014a) Target localization in wireless sensor networks using error correcting codes. IEEE Trans Inf Theory 60(1):697–712. https://doi.org/10.1109/TIT.2013.2289859
34. Vempaty A, Varshney LR, Varshney PK (2014b) Reliable crowdsourcing for multi-class labeling using coding theory. IEEE J Sel Topics Signal Process 8(4):667–679. https://doi.org/10.1109/JSTSP.2014.2316116
35. Vempaty A, Kailkhura B, Varshney PK (2018a) Human-machine inference networks for smart decision making: opportunities and challenges. In: Proceedings of the IEEE international conference on acoustics, speech, and signal processing (ICASSP 2018)
36. Vempaty A, Varshney LR, Koop GJ, Criss AH, Varshney PK (2018b) Experiments and models for decision fusion by humans in inference networks. IEEE Trans Signal Process 66(11):2960–2971. https://doi.org/10.1109/TSP.2017.2784358

37. Vivacqua R, Vassallo R, Martins F (2017) A low cost sensors approach for accurate vehicle localization and autonomous driving application. Sensors 17(10):1–33. https://doi.org/10.3390/s17102359
38. Wang TY, Han YS, Varshney PK, Chen PN (2005) Distributed fault-tolerant classification in wireless sensors networks. IEEE J Sel Areas Commun 23(4):724–734. https://doi.org/10.1109/JSAC.2005.843541
39. White CN, Ratcliff R, Vasey MW, McKoon G (2010) Anxiety enhances threat processing without competition among multiple inputs: a diffusion model analysis. Emotion 10(5):662–677. https://doi.org/10.1037/a0019474
40. Wyner AD (1975) The wire-tap channel. Bell Syst Techn J 54(8):1355–1387. https://doi.org/10.1002/j.1538-7305.1975.tb02040.x

Appendix A

A.1 Proof of Lemma 3.2

A function $f(P_{1,0})$ is quasi-convex if, for some $P^*_{1,0}$, $f(P_{1,0})$ is non-increasing for $P_{1,0} \leq P^*_{1,0}$ and $f(P_{1,0})$ is non-decreasing for $P_{1,0} \geq P^*_{1,0}$. In other words, the lemma is proved if $\frac{dP_E}{dP_{1,0}} \leq 0$ (or $\frac{dP_E}{dP_{1,0}} \geq 0$) for all $P_{1,0}$, or if for some $P^*_{1,0}$, $\frac{dP_E}{dP_{1,0}} \leq 0$ when $P_{1,0} \leq P^*_{1,0}$ and $\frac{dP_E}{dP_{1,0}} \geq 0$ when $P_{1,0} \geq P^*_{1,0}$. First, calculate the partial derivative of P_E with respect to $P_{1,0}$ for an arbitrary K as follows:

$$\frac{dP_E}{dP_{1,0}} = P_0 \frac{dQ_F}{dP_{1,0}} - P_1 \frac{dQ_D}{dP_{1,0}}. \tag{A.1}$$

The derivation of $\frac{dP_E}{dP_{1,0}}$ is given below.

$$\frac{dQ_F}{dP_{1,0}} = \alpha(1 - P_f)N \binom{N-1}{K-1} (\pi_{1,0})^{K-1} (1 - \pi_{1,0})^{N-K}, \tag{A.2}$$

$$\frac{dQ_D}{dP_{1,0}} = \alpha(1 - P_d)N \binom{N-1}{K-1} (\pi_{1,1})^{K-1} (1 - \pi_{1,1})^{N-K}, \tag{A.3}$$

and

$$\frac{dP_E}{dP_{1,0}} = -P_1\alpha(1 - P_d)N \binom{N-1}{K-1} (\pi_{1,1})^{K-1} (1 - \pi_{1,1})^{N-K}$$
$$+ P_0\alpha(1 - P_f)N \binom{N-1}{K-1} (\pi_{1,0})^{K-1} (1 - \pi_{1,0})^{N-K}. \tag{A.4}$$

A. Vempaty et al., *Secure Networked Inference with Unreliable Data Sources*,
https://doi.org/10.1007/978-981-13-2312-6

$\frac{dP_E}{dP_{1,0}}$ given in (A.4) can be reformulated as follows:

$$\frac{dP_E}{dP_{1,0}} = g\left(P_{1,0}, K, \alpha\right)\left(e^{r\left(P_{1,0},K,\alpha\right)} - 1\right), \tag{A.5}$$

where

$$g\left(P_{1,0}, K, \alpha\right) = N\binom{N-1}{K-1}P_1\alpha(1-P_d)(\pi_{1,1})^{K-1}(1-\pi_{1,1})^{N-K} \tag{A.6}$$

and

$$\begin{aligned} r\left(P_{1,0}, K, \alpha\right) &= \ln\left(\frac{P_0}{P_1}\frac{1-P_f}{1-P_d}\left(\frac{\pi_{1,0}}{\pi_{1,1}}\right)^{(K-1)}\left(\frac{1-\pi_{1,0}}{1-\pi_{1,1}}\right)^{(N-K)}\right) \\ &= \ln\frac{P_0}{P_1}\frac{1-P_f}{1-P_d} + (K-1)\ln\frac{\pi_{1,0}}{\pi_{1,1}} + (N-K)\ln\frac{1-\pi_{1,0}}{1-\pi_{1,1}} \end{aligned} \tag{A.7}$$

It can be seen that $g\left(P_{1,0}, K, \alpha\right) \geq 0$ so that the sign of $\frac{dP_E}{dP_{1,0}}$ depends only on the value of $r\left(P_{1,0}, K, \alpha\right)$. To prove that P_E is a quasi-convex function of $P_{1,0}$ when the majority rule K^* is used at the FC, it is sufficient to show that $r\left(P_{1,0}, K^*, \alpha\right)$ is a non-decreasing function. Differentiating $r\left(P_{1,0}, K^*, \alpha\right)$ with respect to $P_{1,0}$, we get

$$\begin{aligned} \frac{dr\left(P_{1,0}, K^*, \alpha\right)}{dP_{1,0}} &= (K^*-1)\left(\frac{\alpha(1-P_f)}{\pi_{1,0}} - \frac{\alpha(1-P_d)}{\pi_{1,1}}\right) + (N-K^*)\left(\frac{\alpha(1-P_d)}{1-\pi_{1,1}} - \frac{\alpha(1-P_f)}{1-\pi_{1,0}}\right) \\ &= (K^*-1)\alpha\left(\frac{1-P_f}{\pi_{1,0}} - \frac{1-P_d}{\pi_{1,1}}\right) - (N-K^*)\alpha\left(\frac{1-P_f}{1-\pi_{1,0}} - \frac{1-P_d}{1-\pi_{1,1}}\right). \end{aligned} \tag{A.8}$$

It can be shown that $\frac{dr\left(P_{1,0}, K^*, \alpha\right)}{dP_{1,0}} > 0$, and this completes the proof. □

A.2 Proof of Lemma 3.3

For a fixed $P_{1,0}$, we have

$$(\pi_{1,0})' = d\pi_{1,0}/dP_{0,1} = \alpha(-P_f). \tag{A.9}$$

For an arbitrary K, we have

$$\frac{dP_E}{dP_{0,1}} = P_1 \alpha P_d N \binom{N-1}{K-1} \left(\pi_{1,1}\right)^{K-1} \left(1-\pi_{1,1}\right)^{N-K}$$

$$- P_0 \alpha P_f N \binom{N-1}{K-1} \left(\pi_{1,0}\right)^{K-1} \left(1-\pi_{1,0}\right)^{N-K}. \tag{A.10}$$

$\frac{dP_E}{dP_{0,1}}$ given in (A.10) can be reformulated as follows:

$$\frac{dP_E}{dP_{0,1}} = g\left(P_{0,1}, K, \alpha\right)\left(e^{r\left(P_{0,1}, K, \alpha\right)} - 1\right), \tag{A.11}$$

where

$$g\left(P_{0,1}, K, \alpha\right) = N \binom{N-1}{K-1} P_0 \alpha P_f (\pi_{1,0})^{K-1} (1-\pi_{1,0})^{N-K} \tag{A.12}$$

and

$$\begin{aligned} r\left(P_{0,1}, K, \alpha\right) &= \ln\left(\frac{P_1}{P_0}\frac{P_d}{P_f}\left(\frac{\pi_{1,1}}{\pi_{1,0}}\right)^{(K-1)}\left(\frac{1-\pi_{1,1}}{1-\pi_{1,0}}\right)^{(N-K)}\right) \\ &= \ln\frac{P_1}{P_0}\frac{P_d}{P_f} + (K-1)\ln\frac{\pi_{1,1}}{\pi_{1,0}} + (N-K)\ln\frac{1-\pi_{1,1}}{1-\pi_{1,0}}. \end{aligned} \tag{A.13}$$

It can be seen that $g\left(P_{0,1}, K, \alpha\right) \geq 0$ such that the sign of $\frac{dP_E}{dP_{0,1}}$ depends on the value of $r\left(P_{0,1}, K, \alpha\right)$. To prove that P_E is a quasi-convex function of $P_{1,0}$ when the majority rule K^* is used at the FC, it is sufficient to show that $r\left(P_{0,1}, K^*, \alpha\right)$ is a non-decreasing function. Differentiating $r\left(P_{0,1}, K^*, \alpha\right)$ with respect to $P_{0,1}$, we get

$$\frac{dr\left(P_{0,1}, K^*, \alpha\right)}{dP_{0,1}} = (K^*-1)\left(\frac{\alpha P_f}{\pi_{1,0}} - \frac{\alpha P_d}{\pi_{1,1}}\right) + (N-K^*)\left(\frac{\alpha P_d}{1-\pi_{1,1}} - \frac{\alpha P_f}{1-\pi_{1,0}}\right) \tag{A.14}$$

$$= (N-K^*)\alpha\left(\frac{P_d}{1-\pi_{1,1}} - \frac{P_f}{1-\pi_{1,0}}\right) - (K^*-1)\alpha\left(\frac{P_d}{\pi_{1,1}} - \frac{P_f}{\pi_{1,0}}\right). \tag{A.15}$$

In the following, we show that

$$\frac{dr\left(P_{0,1}, K^*, \alpha\right)}{dP_{0,1}} > 0, \tag{A.16}$$

i.e., $r\left(P_{0,1}, K^*, \alpha\right)$ is non-decreasing. It is sufficient to show that

$$(N-K^*)\left(\frac{P_d}{1-\pi_{1,1}}-\frac{P_f}{1-\pi_{1,0}}\right) > (K^*-1)\left(\frac{P_d}{\pi_{1,1}}-\frac{P_f}{\pi_{1,0}}\right). \tag{A.17}$$

First, consider the case when there are an even number of nodes in the network and the majority fusion rule is given by $K^* = \frac{N}{2}+1$. Since $0 \le \pi_{1,0} < \pi_{1,1} \le 1$ and $N \ge 2$, we have

$$\begin{aligned}
&\left(1-\frac{2}{N}\right)\frac{\pi_{1,1}\pi_{1,0}}{(1-\pi_{1,1})(1-\pi_{1,0})} > -1\\
\Leftrightarrow &\left(1-\frac{2}{N}\right)\left[\frac{1}{1-\pi_{1,1}}-\frac{1}{1-\pi_{1,0}}\right] > \left[\frac{1}{\pi_{1,1}}-\frac{1}{\pi_{1,0}}\right]\\
\Leftrightarrow &\left[\left(1-\frac{2}{N}\right)\frac{1}{1-\pi_{1,1}}-\frac{1}{\pi_{1,1}}\right] > \left[\left(1-\frac{2}{N}\right)\frac{1}{1-\pi_{1,0}}-\frac{1}{\pi_{1,0}}\right].
\end{aligned} \tag{A.18}$$

Using the fact that $\frac{P_d}{P_f} > 1$, $\pi_{1,1} > \frac{N}{2N-2}$, and $K^* = \frac{N}{2}+1$, (A.18) becomes

$$\begin{aligned}
&\frac{P_d}{P_f}\left[\left(1-\frac{2}{N}\right)\frac{1}{1-\pi_{1,1}}-\frac{1}{\pi_{1,1}}\right] > \left[\left(1-\frac{2}{N}\right)\frac{1}{1-\pi_{1,0}}-\frac{1}{\pi_{1,0}}\right]\\
\Leftrightarrow &\left(1-\frac{2}{N}\right)\frac{P_d}{1-\pi_{1,1}}-\frac{P_d}{\pi_{1,1}} > \left(1-\frac{2}{N}\right)\frac{P_f}{1-\pi_{1,0}}-\frac{P_f}{\pi_{1,0}}\\
\Leftrightarrow &(N-K^*)\left(\frac{P_d}{1-\pi_{1,1}}-\frac{P_f}{1-\pi_{1,0}}\right) > (K^*-1)\left(\frac{P_d}{\pi_{1,1}}-\frac{P_f}{\pi_{1,0}}\right).
\end{aligned} \tag{A.19}$$

Next, consider the case when there are odd number of nodes in the network and the majority fusion rule is given by $K^* = \frac{N+1}{2}$. By using the fact that $\frac{\pi_{1,0}}{\pi_{1,1}} > \frac{P_f}{P_d}$, it can be seen that the right-hand side of (A.19) is nonnegative. Hence, from (A.19), we have

$$\begin{aligned}
&\left(\frac{N}{2}-1\right)\left(\frac{P_d}{1-\pi_{1,1}}-\frac{P_f}{1-\pi_{1,0}}\right) > \frac{N}{2}\left(\frac{P_d}{\pi_{1,1}}-\frac{P_f}{\pi_{1,0}}\right)\\
\Leftrightarrow &\left(\frac{N-1}{2}\right)\left(\frac{P_d}{1-\pi_{1,1}}-\frac{P_f}{1-\pi_{1,0}}\right) > \left(\frac{N-1}{2}\right)\left(\frac{P_d}{1-\pi_{1,1}}-\frac{P_f}{1-\pi_{1,0}}\right)\\
\Leftrightarrow &(N-K^*)\left(\frac{P_d}{1-\pi_{1,1}}-\frac{P_f}{1-\pi_{1,0}}\right) > (K^*-1)\left(\frac{P_d}{\pi_{1,1}}-\frac{P_f}{\pi_{1,0}}\right).
\end{aligned}$$

This completes the proof. □

A.3 Proof of Lemma 3.4

Consider the maximum a posteriori probability (MAP) rule

$$\frac{P(\mathbf{u}|H_1)}{P(\mathbf{u}|H_0)} \underset{H_0}{\overset{H_1}{\gtrless}} \frac{P_0}{P_1}.$$

Since the u_is are independent of each other, the MAP rule simplifies to

$$\prod_{i=1}^{N} \frac{P(u_i|H_1)}{P(u_i|H_0)} \underset{H_0}{\overset{H_1}{\gtrless}} \frac{P_0}{P_1}.$$

Let us assume that K^* out of N nodes send $u_i = 1$. Now, the above equation can be written as

$$\frac{\pi_{1,1}^{K^*}(1-\pi_{1,1})^{N-K^*}}{\pi_{1,0}^{K^*}(1-\pi_{1,0})^{N-K^*}} \underset{H_0}{\overset{H_1}{\gtrless}} \frac{P_0}{P_1}.$$

Taking logarithms on both sides of the above equation, we have

$$K^* \ln \pi_{1,1} + (N-K^*)\ln(1-\pi_{1,1}) - K^* \ln \pi_{1,0} - (N-K^*)\ln(1-\pi_{1,0}) \underset{H_0}{\overset{H_1}{\gtrless}} \ln \frac{P_0}{P_1}$$

$$\Leftrightarrow K^*[\ln(\pi_{1,1}/\pi_{1,0}) + \ln((1-\pi_{1,0})/(1-\pi_{1,1}))] \underset{H_0}{\overset{H_1}{\gtrless}} \ln \frac{P_0}{P_1} + N\ln((1-\pi_{1,0})/(1-\pi_{1,1}))$$

$$\Leftrightarrow K^* \underset{H_0}{\overset{H_1}{\gtrless}} \frac{\ln \frac{P_0}{P_1} + N\ln((1-\pi_{1,0})/(1-\pi_{1,1}))}{[\ln(\pi_{1,1}/\pi_{1,0}) + \ln((1-\pi_{1,0})/(1-\pi_{1,1}))]} \tag{A.20}$$

$$\Leftrightarrow K^* \underset{H_0}{\overset{H_1}{\gtrless}} \frac{\ln\left[(P_0/P_1)\left\{(1-\pi_{1,0})/(1-\pi_{1,1})\right\}^N\right]}{\ln\left[\{\pi_{1,1}(1-\pi_{1,0})\}/\{\pi_{1,0}(1-\pi_{1,1})\}\right]},$$

where (A.20) follows from the fact that, for $\pi_{1,1} > \pi_{1,0}$ or equivalently, $\alpha < \frac{1}{P_{0,1}+P_{1,0}}$, $[\ln(\pi_{1,1}/\pi_{1,0}) + \ln((1-\pi_{1,0})/(1-\pi_{1,1}))] > 0$. □

A.4 Proof of Lemma 3.6

Observe that, for a fixed λ, $P_E(\lceil K^* \rceil)$ is a continuous but not a differentiable function. However, the function is non-differentiable only at a finite number (or countable infinitely number) of points because of the nature of $\lceil K^* \rceil$. Now observe that, for a fixed fusion rule K, $P_E(K)$ is differentiable. Utilizing this fact, to show that the lemma is true, we first find the condition that a fusion rule K should satisfy so that

P_E is a monotonically increasing function of $P_{1,0}$ while keeping $P_{0,1}$ fixed (and vice versa) and later show that $\lceil K^* \rceil$ satisfies this condition. From (A.5), finding those K that satisfy $\frac{dP_E}{dP_{1,0}} > 0$[1] is equivalent to finding those values of K that make

$$
\begin{aligned}
& r\left(P_{1,0}, K, \alpha\right) > 0 \\
\Leftrightarrow\ & \ln \frac{P_0}{P_1} \frac{1 - P_f}{1 - P_d} + (K-1) \ln \frac{\pi_{1,0}}{\pi_{1,1}} + (N-K) \ln \frac{1 - \pi_{1,0}}{1 - \pi_{1,1}} > 0 \\
\Leftrightarrow\ & K < \frac{\ln \frac{P_0}{P_1} + N \ln \frac{(1-\pi_{1,0})}{(1-\pi_{1,1})} + \ln \frac{1 - P_f}{1 - P_d} - \ln \frac{\pi_{1,0}}{\pi_{1,1}}}{\ln\left[\{\pi_{1,1}(1-\pi_{1,0})\}/\{\pi_{1,0}(1-\pi_{1,1})\}\right]}.
\end{aligned}
\tag{A.21}
$$

Similarly, one can find the condition that a fusion rule K should satisfy so that P_E is a monotonically increasing function of $P_{0,1}$ while keeping $P_{1,0}$ fixed. From (A.11), finding those K that satisfy $\frac{dP_E}{dP_{0,1}} > 0$ is equivalent to finding those K that make

$$
\begin{aligned}
& r\left(P_{0,1}, K, \alpha\right) > 0 \\
\Leftrightarrow\ & \ln \frac{P_1}{P_0} \frac{P_d}{P_f} + (K-1) \ln \frac{\pi_{1,1}}{\pi_{1,0}} + (N-K) \ln \frac{1 - \pi_{1,1}}{1 - \pi_{1,0}} > 0 \\
\Leftrightarrow\ & K > \frac{\ln \frac{P_0}{P_1} + N \ln \frac{(1-\pi_{1,0})}{(1-\pi_{1,1})} + \ln \frac{P_f}{P_d} - \ln \frac{\pi_{1,0}}{\pi_{1,1}}}{\ln\left[\{\pi_{1,1}(1-\pi_{1,0})\}/\{\pi_{1,0}(1-\pi_{1,1})\}\right]}.
\end{aligned}
\tag{A.22}
$$

From (A.21) and (A.22), we have

$$
A = \frac{\ln \frac{P_0}{P_1} + N \ln \frac{(1-\pi_{1,0})}{(1-\pi_{1,1})} + \ln \frac{1 - P_f}{1 - P_d} - \ln \frac{\pi_{1,0}}{\pi_{1,1}}}{\ln\left[\{\pi_{1,1}(1-\pi_{1,0})\}/\{\pi_{1,0}(1-\pi_{1,1})\}\right]} > K > \frac{\ln \frac{P_0}{P_1} + N \ln \frac{(1-\pi_{1,0})}{(1-\pi_{1,1})} + \ln \frac{P_f}{P_d} - \ln \frac{\pi_{1,0}}{\pi_{1,1}}}{\ln\left[\{\pi_{1,1}(1-\pi_{1,0})\}/\{\pi_{1,0}(1-\pi_{1,1})\}\right]} = B.
\tag{A.23}
$$

Next, let us show that the optimal fusion rule $\lceil K^* \rceil$ given in (3.20) is within the region (A, B). First, we prove that $\lceil K^* \rceil > B$ by showing $K^* > B$. Comparing K^* given in (3.20) with B, $K^* > B$ iff

$$
0 > \ln \frac{P_f}{P_d} - \ln \frac{\pi_{1,0}}{\pi_{1,1}}.
\tag{A.24}
$$

[1]Observe that, for $\alpha < 0.5$, the function $g\left(P_{1,0}, K^*, \alpha\right) = 0$ (as given in Eq. A.6) only under extreme conditions (i.e., $P_1 = 0$ or $P_d = 0$ or $P_d = 1$). Ignoring these extreme conditions, we have $g\left(P_{1,0}, K^*, \alpha\right) > 0$.

Since $P_d > P_f$, to prove (A.24) we start from the inequality

$$
\begin{aligned}
&\frac{(1-P_d)}{P_d} < \frac{(1-P_f)}{P_f} \\
\Leftrightarrow\ &\frac{\alpha P_{1,0}(1-P_d) + P_d(1-P_{0,1}\alpha)}{P_d} < \frac{\alpha P_{1,0}(1-P_f) + P_f(1-P_{0,1}\alpha)}{P_f} \\
\Leftrightarrow\ &\frac{\pi_{1,1}}{P_d} < \frac{\pi_{1,0}}{P_f} \\
\Leftrightarrow\ &0 > \ln\frac{P_f}{P_d} - \ln\frac{\pi_{1,0}}{\pi_{1,1}}.
\end{aligned}
$$

Now, we show that $A > \lceil K^* \rceil$. Observe that,

$$
\begin{aligned}
&A > \lceil K^* \rceil \\
\Leftrightarrow\ &\frac{\ln\dfrac{1-P_f}{1-P_d} - \ln\dfrac{\pi_{1,0}}{\pi_{1,1}}}{\ln\left[\{\pi_{1,1}(1-\pi_{1,0})\}/\{\pi_{1,0}(1-\pi_{1,1})\}\right]} > \lceil K^* \rceil - K^*.
\end{aligned}
$$

Hence, it is sufficient to show that

$$
\frac{\ln\dfrac{1-P_f}{1-P_d} - \ln\dfrac{\pi_{1,0}}{\pi_{1,1}}}{\ln\left[\{\pi_{1,1}(1-\pi_{1,0})\}/\{\pi_{1,0}(1-\pi_{1,1})\}\right]} > 1 > \lceil K^* \rceil - K^*.
$$

$1 > \lceil K^* \rceil - K^*$ is true from the property of the ceiling function. We have

$$
\begin{aligned}
&\frac{1-P_f}{1-P_d} > \frac{1-\pi_{1,0}}{1-\pi_{1,1}} \\
\Leftrightarrow\ &\ln\frac{1-P_f}{1-P_d} > \ln\frac{1-\pi_{1,0}}{1-\pi_{1,1}} \\
\Leftrightarrow\ &\ln\frac{1-P_f}{1-P_d} - \ln\frac{\pi_{1,0}}{\pi_{1,1}} > \ln\left[\{\pi_{1,1}(1-\pi_{1,0})\}/\{\pi_{1,0}(1-\pi_{1,1})\}\right] \\
\Leftrightarrow\ &\frac{\ln\dfrac{1-P_f}{1-P_d} - \ln\dfrac{\pi_{1,0}}{\pi_{1,1}}}{\ln\left[\{\pi_{1,1}(1-\pi_{1,0})\}/\{\pi_{1,0}(1-\pi_{1,1})\}\right]} > 1
\end{aligned}
$$

which completes the proof. □

A.5 Proof of Proposition 3.1

The blinding is possible when

$$P(z = j|H_0) = P(z = j|H_1) \quad \forall j \in \{0, 1\}. \tag{A.25}$$

Substituting (3.54) and (3.55) in (A.25) and after simplification, the condition to make the $KLD = 0$ for a K-level network can be expressed as

$$P^B_{j,1} - P^B_{j,0} = \frac{\sum_{k=1}^{K}[\beta_k(1 - \sum_{i=1}^{k}\alpha_i)]}{\sum_{k=1}^{K}[\beta_k(\sum_{i=1}^{k}\alpha_i)]} \frac{P^H_d - P^H_{fa}}{P^B_d - P^B_{fa}} (P^H_{j,0} - P^H_{j,1}). \tag{A.26}$$

From (3.49) to (3.52), we have

$$P^B_{0,1} - P^B_{0,0} = \frac{\sum_{k=1}^{K}[\beta_k(1 - \sum_{i=1}^{k}\alpha_i)]}{\sum_{k=1}^{K}[\beta_k(\sum_{i=1}^{k}\alpha_i)]} \frac{P^H_d - P^H_{fa}}{P^B_d - P^B_{fa}} = -(P^B_{1,1} - P^B_{1,0}). \tag{A.27}$$

Hence, the attacker can degrade detection performance by intelligently choosing $(P^B_{0,1}, P^B_{1,0})$, which are dependent on α_k, for $k = 1, \cdots, K$. Observe that,

$$0 \leq P^B_{0,1} - P^B_{0,0}$$

since $\sum_{i=1}^{k}\alpha_i \leq 1$ for $k \leq K$. To make $KLD = 0$, we must have

$$P^B_{0,1} - P^B_{0,0} \leq 1$$

such that $(P^B_{j,1}, P^B_{j,0})$ becomes a valid probability mass function. Notice that, when $P^B_{0,1} - P^B_{0,0} > 1$, there does not exist any attacking probability distribution $(P^B_{j,1}, P^B_{j,0})$ that can make $KLD = 0$. In the case of $P^B_{0,1} - P^B_{0,0} = 1$, there exists a unique solution $(P^B_{1,1}, P^B_{1,0}) = (0, 1)$ that can make $KLD = 0$. For the $P^B_{0,1} - P^B_{0,0} < 1$ case, there exist an infinite number of attacking probability distributions $(P^B_{j,1}, P^B_{j,0})$ which can make $KLD = 0$.

By further assuming that the honest and Byzantine nodes are identical in terms of their detection performance, i.e., $P^H_d = P^B_d$ and $P^H_{fa} = P^B_{fa}$, the above condition to blind the FC reduces to

$$\frac{\sum_{k=1}^{K}[\beta_k(1 - \sum_{i=1}^{k}\alpha_i)]}{\sum_{k=1}^{K}[\beta_k(\sum_{i=1}^{k}\alpha_i)]} \leq 1$$

which is equivalent to

$$\sum_{k=1}^{K}[\beta_k(1-2(\sum_{i=1}^{k}\alpha_i))] \le 0. \tag{A.28}$$

Recall that $\alpha_k = \frac{B_k}{N_k}$ and $\beta_k = \frac{N_k}{\sum_{i=1}^{K} N_i}$. Substituting α_k and β_k in (A.28) and simplifying the expressions, we have the desired result. □

A.6 Proof of Lemma 3.9

Let us denote, $P(z=1|H_1) = \pi_{1,1}$, $P(z=1|H_0) = \pi_{1,0}$ and $t = \sum_{k=1}^{K}\beta_k\sum_{i=1}^{k}\alpha_i$. Notice that, in the region where the attacker cannot blind the FC, the parameter $t < 0.5$. To prove the lemma, we first show that any positive deviation $\varepsilon \in (0, p]$ in flipping probabilities $(P_{1,0}^B, P_{0,1}^B) = (p, p-\varepsilon)$ will result in an increase in the KLD. After plugging in $(P_{1,0}^B, P_{0,1}^B) = (p, p-\varepsilon)$ in (3.54) and (3.55), we get

$$\pi_{1,1} = t(p - P_d(2p-\varepsilon)) + P_d \tag{A.29}$$

$$\pi_{1,0} = t(p - P_{fa}(2p-\varepsilon)) + P_{fa}. \tag{A.30}$$

Now, we show that the KLD, D, as given in (3.53) is a monotonically increasing function of the parameter ε, or in other words, $\frac{dD}{d\varepsilon} > 0$.

$$\begin{aligned}\frac{dD}{d\varepsilon} &= \pi_{1,1}\left(\frac{\pi'_{1,1}}{\pi_{1,1}} - \frac{\pi'_{1,0}}{\pi_{1,0}}\right) + \pi'_{1,1}\log\frac{\pi_{1,1}}{\pi_{1,0}} \\ &+ (1-\pi_{1,1})\left(\frac{\pi'_{1,0}}{1-\pi_{1,0}} - \frac{\pi'_{1,1}}{1-\pi_{1,1}}\right) - \pi'_{1,1}\log\frac{1-\pi_{1,1}}{1-\pi_{1,0}}\end{aligned} \tag{A.31}$$

where $\frac{d\pi_{1,1}}{d\varepsilon} = \pi'_{1,1} = tP_d$ and $\frac{d\pi_{1,0}}{d\varepsilon} = \pi'_{1,0} = tP_{fa}$ and t is the fraction of covered nodes by the Byzantines. After rearranging the terms in the above equation, the condition $\frac{dD}{d\varepsilon} > 0$ becomes

$$\frac{1-\pi_{1,1}}{1-\pi_{1,0}} + \frac{P_d}{P_{fa}}\log\frac{\pi_{1,1}}{\pi_{1,0}} > \frac{\pi_{1,1}}{\pi_{1,0}} + \frac{P_d}{P_{fa}}\log\frac{1-\pi_{1,1}}{1-\pi_{1,0}}. \tag{A.32}$$

Since $P_d > P_{fa}$ and $t < 0.5$, $\pi_{1,1} > \pi_{1,0}$. It can also be proved that $\frac{P_{fa}}{P_d}\frac{\pi_{1,1}}{\pi_{1,0}} < 1$. Hence, we have

$$
\begin{aligned}
&1+(\pi_{1,1}-\pi_{1,0}) > \frac{P_{fa}}{P_d}\frac{\pi_{1,1}}{\pi_{1,0}} \\
&\Leftrightarrow (\pi_{1,1}-\pi_{1,0})\left[\frac{1+(\pi_{1,1}-\pi_{1,0})}{\pi_{1,1}(1-\pi_{1,0})}\right] > \frac{P_{fa}}{P_d}\frac{\pi_{1,1}}{\pi_{1,0}}\left[\frac{\pi_{1,1}-\pi_{1,0}}{\pi_{1,1}(1-\pi_{1,0})}\right] \\
&\Leftrightarrow \left[\frac{1-\pi_{1,0}-(1-\pi_{1,1})}{1-\pi_{1,0}}+\frac{(\pi_{1,1}-\pi_{1,0})}{\pi_{1,1}}\right] > \frac{P_{fa}}{P_d}\left[\frac{\pi_{1,1}}{\pi_{1,0}}-\frac{1-\pi_{1,1}}{1-\pi_{1,0}}\right] \\
&\Leftrightarrow \frac{1-\pi_{1,1}}{1-\pi_{1,0}}+\frac{P_d}{P_{fa}}\left(1-\frac{\pi_{1,0}}{\pi_{1,1}}\right) > \frac{\pi_{1,1}}{\pi_{1,0}}+\frac{P_d}{P_{fa}}\left(\frac{1-\pi_{1,1}}{1-\pi_{1,0}}-1\right). \qquad (A.33)
\end{aligned}
$$

To prove that (A.32) is true, we apply the logarithm inequality $(x-1) \geq \log x \geq \frac{x-1}{x}$, for $x > 0$ to (A.33). First, let us assume that $x = \frac{\pi_{1,1}}{\pi_{1,0}}$. Now, using the logarithm inequality, we can show that $\log\frac{\pi_{1,1}}{\pi_{1,0}} \geq 1-\frac{\pi_{1,0}}{\pi_{1,1}}$. Next, let us assume that $x = \frac{1-\pi_{1,1}}{1-\pi_{1,0}}$. Again, using the logarithm inequality, it can be shown that $\left[\frac{1-\pi_{1,1}}{1-\pi_{1,0}}-1\right] \geq \log\frac{1-\pi_{1,1}}{1-\pi_{1,0}}$. Using these results and (A.33), one can prove that condition (A.32) is true.

Similarly, we can show that any nonzero deviation $\varepsilon \in (0, p]$ in flipping probabilities $(P^B_{1,0}, P^B_{0,1}) = (p-\varepsilon, p)$ will result in an increase in the KLD, i.e., $\frac{dD}{d\varepsilon} > 0$, or

$$
\frac{\pi_{1,1}}{\pi_{1,0}}+\frac{1-P_d}{1-P_{fa}}\log\frac{1-\pi_{1,1}}{1-\pi_{1,0}} > \frac{1-\pi_{1,1}}{1-\pi_{1,0}}+\frac{1-P_d}{1-P_{fa}}\log\frac{\pi_{1,1}}{\pi_{1,0}}. \qquad (A.34)
$$

Since $P_d > P_{fa}$ and $t < 0.5$, $\pi_{1,1} > \pi_{1,0}$. It can also be proved that $\frac{1-\pi_{1,1}}{1-\pi_{1,0}} > \frac{1-P_d}{1-P_{fa}}$. Hence, we have

$$
\frac{1-\pi_{1,1}}{1-\pi_{1,0}} > \frac{1-P_d}{1-P_{fa}}[1-(\pi_{1,1}-\pi_{1,0})] \qquad (A.35)
$$

$$
\begin{aligned}
&\Leftrightarrow \frac{1-\pi_{1,1}}{\pi_{1,0}(1-\pi_{1,0})} > \frac{1-P_d}{1-P_{fa}}\left[\frac{1-(\pi_{1,1}-\pi_{1,0})}{\pi_{1,0}}\right] \\
&\Leftrightarrow \frac{1}{\pi_{1,0}(1-\pi_{1,0})} > \frac{1-P_d}{1-P_{fa}}\left[\frac{1-(\pi_{1,1}-\pi_{1,0})}{\pi_{1,0}(1-\pi_{1,1})}\right] \\
&\Leftrightarrow \frac{1}{\pi_{1,1}-\pi_{1,0}}\left[\frac{\pi_{1,1}}{\pi_{1,0}}-\frac{1-\pi_{1,1}}{1-\pi_{1,0}}\right] > \frac{1-P_d}{1-P_{fa}}\left[\frac{1}{\pi_{1,0}}+\frac{1}{1-\pi_{1,1}}\right] \qquad (A.36) \\
&\Leftrightarrow \frac{\pi_{1,1}}{\pi_{1,0}}-\frac{1-\pi_{1,1}}{1-\pi_{1,0}} > \frac{1-P_d}{1-P_{fa}}\left[\frac{\pi_{1,1}-\pi_{1,0}}{\pi_{1,0}}+\frac{\pi_{1,1}-\pi_{1,0}}{1-\pi_{1,1}}\right] \qquad (A.37) \\
&\Leftrightarrow \frac{\pi_{1,1}}{\pi_{1,0}}+\frac{1-P_d}{1-P_{fa}}\left[1-\frac{1-\pi_{1,0}}{1-\pi_{1,1}}\right] > \frac{1-\pi_{1,1}}{1-\pi_{1,0}}+\frac{1-P_d}{1-P_{fa}}\left[\frac{\pi_{1,1}}{\pi_{1,0}}-1\right]. \qquad (A.38)
\end{aligned}
$$

To prove that (A.34) is true, we apply the logarithm inequality $(x-1) \geq \log x \geq \frac{x-1}{x}$, for $x > 0$ to (A.38). First, let us assume that $x = \frac{1-\pi_{1,1}}{1-\pi_{1,0}}$. Now, using the logarithm inequality, we can show that $\log \frac{1-\pi_{1,1}}{1-\pi_{1,0}} \geq 1 - \frac{1-\pi_{1,0}}{1-\pi_{1,1}}$. Next, let us assume that $x = \frac{\pi_{1,1}}{\pi_{1,0}}$. Again, using the logarithm inequality, it can be shown that $\left[\frac{\pi_{1,1}}{\pi_{1,0}} - 1\right] \geq \log \frac{\pi_{1,1}}{\pi_{1,0}}$. Using these results and (A.38), one can prove that condition (A.34) is true. Condition (A.32) and (A.34) imply that any nonzero deviation $\varepsilon_i \in (0, p]$ in flipping probabilities $(P^B_{0,1}, P^B_{1,0}) = (p - \varepsilon_1, p - \varepsilon_2)$ will result in an increase in the KLD. □

A.7 Proof of Theorem 3.4

Observe that, in the region where the attacker cannot blind the FC, the optimal strategy comprises of symmetric flipping probabilities $(P^B_{0,1} = P^B_{1,0} = p)$. The proof is complete if we show that KLD, D, is a monotonically decreasing function of the flipping probability p.

Let us denote, $P(z = 1|H_1) = \pi_{1,1}$ and $P(z = 1|H_0) = \pi_{1,0}$. After plugging in $(P^B_{0,1}, P^B_{1,0}) = (p, p)$ in (3.54) and (3.55), we get

$$\pi_{1,1} = t(p - P_d(2p)) + P_d \tag{A.39}$$

$$\pi_{1,0} = t(p - P_{fa}(2p)) + P_{fa}. \tag{A.40}$$

Now we show that the KLD, D, as given in (3.53) is a monotonically decreasing function of the parameter p, or in other words, $\frac{dD}{dp} < 0$. After plugging in $\pi'_{1,1} = t(1 - 2P_d)$ and $\pi'_{1,0} = t(1 - 2P_{fa})$ in the expression of $\frac{dD}{dp}$ and rearranging the terms, the condition $\frac{dD}{dp} < 0$ becomes

$$(1 - 2P_{fa})\left(\frac{1-\pi_{1,1}}{1-\pi_{1,0}} - \frac{\pi_{1,1}}{\pi_{1,0}}\right) + (1 - 2P_d)\log\left(\frac{1-\pi_{1,0}}{1-\pi_{1,1}}\frac{\pi_{1,1}}{\pi_{1,0}}\right) < 0 \tag{A.41}$$

Since $P_d > P_{fa}$ and $t < 0.5$, we have $\pi_{1,1} > \pi_{1,0}$. Now, using the fact that $\frac{1-P_d}{1-P_{fa}} > \frac{1-2P_d}{1-2P_{fa}}$ and (A.36), we have

$$\frac{1}{\pi_{1,1}-\pi_{1,0}}\left[\frac{\pi_{1,1}}{\pi_{1,0}}-\frac{1-\pi_{1,1}}{1-\pi_{1,0}}\right] > \frac{1-2P_d}{1-2P_{fa}}\left[\frac{1}{\pi_{1,0}}+\frac{1}{1-\pi_{1,1}}\right]$$
$$\Leftrightarrow \frac{\pi_{1,1}}{\pi_{1,0}}+\frac{1-2P_d}{1-2P_{fa}}\left[1-\frac{1-\pi_{1,0}}{1-\pi_{1,1}}\right] > \frac{1-\pi_{1,1}}{1-\pi_{1,0}}+\frac{1-2P_d}{1-2P_{fa}}\left[\frac{\pi_{1,1}}{\pi_{1,0}}-1\right]. \quad (A.42)$$

Applying the logarithm inequality $(x-1) \geq \log x \geq \frac{x-1}{x}$, for $x > 0$ to (A.42), one can prove that (A.41) is true. □

A.8 Proof of Lemma 3.10

Let us denote the optimal attack configuration for a K-level *perfect a-ary* tree topology $T(K, a)$ by $\{B_k^1\}_{k=1}^K$ and the optimal attack configuration for a *perfect a-ary* tree topology with $K+1$ levels by $\{B_k^2\}_{k=1}^{K+1}$ given the cost budget $C_{budget}^{attacker}$. To prove the lemma, it is sufficient to show that

$$\frac{\sum_{k=1}^{K+1} P_k^2 B_k^2}{\sum_{k=1}^{K+1} N_k} \geq \frac{\sum_{k=1}^{K} P_k^2 B_k^1}{\sum_{k=1}^{K+1} N_k} \geq \frac{\sum_{k=1}^{K} P_k^1 B_k^1}{\sum_{k=1}^{K} N_k}, \quad (A.43)$$

where P_k^1 is the profit of attacking a node at level k in a K-level *perfect a-ary* tree topology and P_k^2 is the profit of attacking a node at level k in a $K+1$-level *perfect a-ary* tree topology.

First inequality in (A.43) follows due to the fact that $\{B_k^1\}_{k=1}^K$ may not be the optimal attack configuration for topology $T(K+1, a)$. To prove the second inequality observe that an increase in the value of parameter K results in an increase in both the denominator (number of nodes in the network) and the numerator (fraction of covered nodes). Using this fact, let us denote

$$\frac{\sum_{k=1}^{K} P_k^2 B_k^1}{\sum_{k=1}^{K+1} N_k} = \frac{x+x_1}{y+y_1} \quad (A.44)$$

with $x = \sum_{k=1}^{K} P_k^1 B_k^1$ with $P_k^1 = \frac{a^{K-k+1}-1}{a-1}$, $y = \sum_{k=1}^{K} N_k = \frac{a(a^K-1)}{a-1}$, $x_1 = \sum_{k=1}^{K}(B_k^1 a^{K-k+1})$ is the increase in the profit by adding one more level to the topology and $y_1 = a^{K+1}$ is the increase in the number of nodes in the network by adding one more level to the topology.

Note that $\frac{x+x_1}{y+y_1} > \frac{x}{y}$ if and only if

$$\frac{x}{y} < \frac{x_1}{y_1} \quad (A.45)$$

where x, y, x_1, and y_1 are positive values. Hence, it is sufficient to prove that

$$\frac{a^{K+1}\sum_{k=1}^{K}\left(\frac{B_k^1}{a^k}\right)-\sum_{k=1}^{K}B_k^1}{a(a^K-1)} \le \frac{\sum_{k=1}^{K}(B_k^1 a^{K-k+1})}{a^{K+1}}.$$

The above equation can be further simplified to

$$\sum_{k=1}^{K}\left(\frac{B_k^1}{a^k}\right) \le \sum_{k=1}^{K}\left(\frac{B_k^1}{a}\right)$$

which is true for all $K \ge 1$. □

A.9 Proof of Lemma 3.11

As before, let us denote the optimal attack configuration for a K-level *perfect a-ary* tree topology $T(K,\ a)$ by $\{B_k^1\}_{k=1}^K$ and the optimal attack configuration for a *perfect (a+1)-ary* tree topology $T(K, a+1)$ by $\{B_k^2\}_{k=1}^K$ given the cost budget $C_{budget}^{attacker}$. To prove the lemma, it is sufficient to show that

$$\frac{\sum_{k=1}^{K} P_k^2 B_k^2}{\sum_{k=1}^{K} N_k^2} < \frac{\sum_{k=1}^{K} P_k^1 B_k^2}{\sum_{k=1}^{K} N_k^1} \le \frac{\sum_{k=1}^{K} P_k^1 B_k^1}{\sum_{k=1}^{K} N_k^1} \quad \text{(A.46)}$$

where N_k^1 is the number of nodes at level k in $T(K, a)$, N_k^2 is the number of nodes at level k in $T(K, a+1)$, P_k^1 is the profit of attacking a node at level k in $T(K, a)$, and P_k^2 is the profit of attacking a node at level k in $T(K, a+1)$. Observe that an interpretation of (A.46) is that the attacker is using the attack configuration $\{B_k^2\}_{k=1}^K$ to attack $T(K,\ a)$. However, one might suspect that the set $\{B_k^2\}_{k=1}^{k=K}$ is not a valid solution. More specifically, the set $\{B_k^2\}_{k=1}^{k=K}$ is not a valid solution in the following two cases:

1. $min(B_k^2,\ N_k^1) = N_k^1$ *for any* k: For example, if $N_1^1 = 4$ for $T(K, 4)$ and $B_1^2 = 5$ for $T(K, 5)$, then it will not be possible for the attacker to attack 5 nodes at level 1 in $T(K, 4)$ because the total number of nodes at level 1 is 4. In this case, $\{B_k^2\}_{k=1}^K$ might not be a valid attack configuration for the tree $T(K, a)$.
2. $\{B_k^2\}_{k=1}^{k=K}$ *is an overlapping set*[2] *for* $T(K, a)$: For example, for $T(2, 3)$ if $B_1^2 = 2$ and $B_2^2 = 4$, then B_1^2 and B_2^2 are overlapping. In this case, $\{B_k^2\}_{k=1}^K$ might not be a valid attack configuration for the tree $T(K, a)$.

[2]We call B_k and B_{k+x} are overlapping, if the summation of B_k^{k+x} and B_{k+x} is greater than N_{k+x}, where B_k^{k+x} is the number of nodes covered by the attack configuration B_k at level $k+x$. In a non-overlapping case, the attacker can always arrange nodes $\{B_k\}_{k=1}^K$ such that each path in the network has at most one Byzantine.

However, both of the above conditions imply that the attacker can blind the network with $C_{budget}^{attacker}$, which cannot be true for $a \geq a_{min}$, and therefore, $\{B_k^2\}_{k=1}^K$ will indeed be a valid solution. Therefore, (A.46) is sufficient to prove the lemma.

Notice that the second inequality in (A.46) follows due to the fact that $\{B_k^2\}_{k=1}^K$ may not be the optimal attack configuration for topology $T(K, a)$. To prove the first inequality in (A.46), we first consider the case where attack configuration $\{B_k^2\}_{k=1}^{k=K}$ contains only one node, i.e., $B_k^2 = 1$ for some k, and show that $\frac{P_k^2}{\sum_{k=1}^K N_k^2} < \frac{P_k^1}{\sum_{k=1}^K N_k^1}$. Substituting $P_k^1 = \dfrac{a^{K-k+1} - 1}{a-1}$ for some k and $\sum_{k=1}^K N_k^1 = \dfrac{a(a^K - 1)}{a-1}$ in the left-side inequality of (A.46), we have

$$\frac{(a)^{K-k+1} - 1}{(a)((a)^K - 1)} > \frac{(a+1)^{K-k+1} - 1}{(a+1)((a+1)^K - 1)}.$$

After some simplification, the above condition becomes

$$\begin{aligned}(a+1)^{K+1}[(a)^{K-k+1} - 1] - (a)^{K+1}[(a+1)^{K-k+1} - 1] \\ + (a)[(a+1)^{K-k+1} - 1] - (a+1)[(a)^{K-k+1} - 1] > 0.\end{aligned} \quad \text{(A.47)}$$

It can be shown that

$$(a)[(a+1)^{K-k+1} - 1] - (a+1)[(a)^{K-k+1} - 1] > 0 \quad \text{(A.48)}$$

and

$$(a+1)^{K+1}[(a)^{K-k+1} - 1] - (a)^{K+1}[(a+1)^{K-k+1} - 1] \geq 0. \quad \text{(A.49)}$$

From (A.49) and (A.48), condition (A.47) holds.

Since it is proved that

$$\frac{P_k^2}{\sum_{k=1}^K N_k^2} < \frac{P_k^1}{\sum_{k=1}^K N_k^1} \text{ for all } 1 \leq k \leq K,$$

to generalize the proof for any arbitrary attack configuration $\{B_k^2\}_{k=1}^K$ we multiply both sides of the above inequality with B_k^2 and sum it over all $1 \leq k \leq K$ inequalities. Now, we have

$$\frac{\sum_{k=1}^K P_k^2 B_k^2}{\sum_{k=1}^K N_k^2} < \frac{\sum_{k=1}^K P_k^1 B_k^2}{\sum_{k=1}^K N_k^1}.$$

□

A.10 Proof of Lemma 3.12

Since the KLD is always nonnegative, Byzantines attempt to choose $P(z_i = j|H_0, k)$ and $P(z_i = j|H_1, k)$ such that $D_k = 0,\ \forall k$. This is possible when

$$P(z_i = j|H_0, k) = P(z_i = j|H_1, k) \quad \forall j \in \{0, 1\},\ \forall k. \tag{A.50}$$

Notice that, $\pi^k_{j,0} = P(z_i = j|H_0, k)$ and $\pi^k_{j,1} = P(z_i = j|H_1, k)$ can be expressed as

$$\pi^k_{1,0} = \beta^k_{1,0}(1 - P^k_{fa}) + (1 - \beta^k_{0,1})P^k_{fa} \tag{A.51}$$
$$\pi^k_{1,1} = \beta^k_{1,0}(1 - P^k_d) + (1 - \beta^k_{0,1})P^k_d. \tag{A.52}$$

with $\beta^k_{1,0} = \sum_{j=1}^k \alpha_j P^j_{1,0}$ and $\beta^k_{0,1} = \sum_{j=1}^k \alpha_j P^j_{0,1}$. Substituting (A.51) and (A.52) in (A.50) and after simplification, the condition to make the $D = 0$ for a K-level network becomes $\sum_{j=1}^k \alpha_j (P^j_{1,0} + P^j_{0,1}) = 1,\ \forall k$. Notice that, when $\sum_{j=1}^k \alpha_j < 0.5$, there does not exist any attacking probability distribution $(P^j_{0,1}, P^j_{1,0})$ that can make $D_k = 0$, and, therefore, the KLD cannot be made zero. In the case of $\sum_{j=1}^k \alpha_j = 0.5$, there exists a unique solution $(P^j_{0,0}, P^j_{1,0}) = (1, 1),\ \forall j$ that can make $D_k = 0,\ \forall k$. For the $\sum_{j=1}^k \alpha_j > 0.5$ case, there exist infinitely many attacking probability distributions $(P^j_{0,1}, P^j_{1,0})$ which can make $D_k = 0,\ \forall k$. Now, the proof follows from the fact that the condition $\sum_{j=1}^k \alpha_j = 0.5,\ \forall k$, is equivalent to $\alpha_1 = 0.5,\ \alpha_k = 0,\ \forall k = 2, \cdots, K$. □

A.11 Proof of Theorem 3.5

Observe that, in the region where the attacker cannot make $D_k = 0$, the optimal strategy comprises of symmetric flipping probabilities ($P^k_{0,1} = P^k_{1,0} = p$). The proof is complete if we show that D_k is a monotonically decreasing function of the flipping probability p.

After plugging in $(P^k_{0,1}, P^k_{1,0}) = (p, p)$ in (A.51) and (A.52), we get

$$\pi^k_{1,1} = [\beta^{k-1}_{1,0}(1 - P^k_d) + (1 - \beta^{k-1}_{0,1})P^k_d] + [\alpha_k(p - P^k_d(2p)) + P^k_d] \tag{A.53}$$
$$\pi^k_{1,0} = [\beta^{k-1}_{1,0}(1 - P^k_{fa}) + (1 - \beta^{k-1}_{0,1})P^k_{fa}] + [\alpha_k(p - P^k_{fa}(2p)) + P^k_{fa}]. \tag{A.54}$$

Now, we show that D_k is a monotonically decreasing function of the parameter p, or in other words, $\dfrac{dD_k}{dp} < 0$. After plugging in $\pi^{k'}_{1,1} = \alpha_k(1 - 2P^k_d)$ and $\pi^{k'}_{1,0} =$

$\alpha_k(1-2P_{fa}^k)$ in the expression of $\dfrac{dD_k}{dp}$ and rearranging the terms, the condition $\dfrac{dD_k}{dp} < 0$ becomes

$$(1-2P_d^k)\left(\frac{1-\pi_{1,0}^k}{1-\pi_{1,1}^k}-\frac{\pi_{1,0}^k}{\pi_{1,1}^k}\right)+(1-2P_{fa}^k)\log\left(\frac{1-\pi_{1,1}^k}{1-\pi_{1,0}^k}\frac{\pi_{1,0}^k}{\pi_{1,1}^k}\right)<0 \quad \text{(A.55)}$$

Since $P_d^k > P_{fa}^k$ and $\beta_{\bar{x},x}^k < 0.5$, we have $\pi_{1,1}^k > \pi_{1,0}^k$. Now, using the fact that $\dfrac{1-P_d^k}{1-P_{fa}^k} > \dfrac{1-2P_d^k}{1-2P_{fa}^k}$ and (A.36), we have

$$\frac{1-2P_d^k}{1-2P_{fa}^k}\left[\frac{1-\pi_{1,0}^k}{1-\pi_{1,1}^k}-\frac{\pi_{1,0}^k}{\pi_{1,1}^k}\right] < (\pi_{1,1}^k-\pi_{1,0}^k)\left[\frac{1}{\pi_{1,1}^k}+\frac{1}{1-\pi_{1,0}^k}\right] \quad \text{(A.56)}$$

$$\Leftrightarrow \frac{1-2P_d^k}{1-2P_{fa}^k}\left[\frac{1-\pi_{1,0}^k}{1-\pi_{1,1}^k}-\frac{\pi_{1,0}^k}{\pi_{1,1}^k}\right]+\left[\frac{\pi_{1,0}^k}{\pi_{1,1}^k}-1\right] < 1-\frac{1-\pi_{1,1}^k}{1-\pi_{1,0}^k}. \quad \text{(A.57)}$$

Applying the logarithm inequality $(x-1) \geq \log x \geq \dfrac{x-1}{x}$, for $x > 0$ to (A.57), one can prove that (A.55) is true. □

A.12 Proof of Lemma 3.14

The local test statistic Y_i has the mean

$$mean_i = \begin{cases} M\sigma_i^2 \text{ if } H_0 \\ (M+\eta_i)\sigma_i^2 \text{ if } H_1 \end{cases}$$

and the variance

$$Var_i = \begin{cases} 2M\sigma_i^4 \text{ if } H_0 \\ 2(M+2\eta_i)\sigma_i^4 \text{ if } H_1. \end{cases}$$

The goal of Byzantine nodes is to make the deflection coefficient as small as possible. Since the deflection coefficient is always nonnegative, the Byzantines seek to make $\mathscr{D}(\Lambda) = \dfrac{(\mu_1-\mu_0)^2}{\sigma_{(0)}^2} = 0$. The conditional mean $\mu_k = \mathbb{E}[\Lambda|H_k]$ and conditional variance $\sigma_{(0)}^2 = \mathbb{E}[(\Lambda-\mu_0)^2|H_0]$ of the global test statistic, $\Lambda = (\sum_{i=1}^{N_1}\tilde{w}_i\tilde{Y}_i + \sum_{i=N_1+1}^{N} w_iY_i)/(\text{sum}(w))$, can be computed and are given by (3.75),

(3.76), and (3.77), respectively. After substituting values from (3.75), (3.76), and (3.77), the condition to make $\mathscr{D}(\Lambda) = 0$ becomes

$$\sum_{i=1}^{N_1} \tilde{w}_i (2P_i \Delta_i - \eta_i \sigma_i^2) = \sum_{i=N_1+1}^{N} w_i \eta_i \sigma_i^2$$

□

A.13 Proof of Proposition 4.2

The probability of a sensor sending a bit value 1 is

$$P(u = 1|a) = Q\left(\frac{\eta(\hat{a}) - a}{\sigma}\right). \tag{A.58}$$

The data's contribution to the posterior Fisher information is given by

$$F = -E\left[\frac{\partial^2 \ln P(u|a)}{\partial^2 a}\right], \tag{A.59}$$

where $\ln p(u|a) = (1-u)\ln(1 - P(u = 1|a)) + u \ln P(u = 1|a)$. Let $P_1 = P(u = 1|a)$ and $P_0 = 1 - P_1$. Then

$$\frac{\partial^2 \ln p(u|a)}{\partial^2 a} = -\frac{(1-u)}{P_0^2}\left(\frac{\partial P_0}{\partial a}\right)^2 + \frac{(1-u)}{P_0}\frac{\partial^2 P_0}{\partial^2 a} - \frac{u}{P_1^2}\left(\frac{\partial P_1}{\partial a}\right)^2 + \frac{u}{P_1}\frac{\partial^2 P_1}{\partial^2 a}$$

and

$$E\left[\frac{\partial^2 \ln p(u|a)}{\partial^2 a}\right] = -\frac{1}{P_0}\left(\frac{\partial P_0}{\partial a}\right)^2 - \frac{1}{P_1}\left(\frac{\partial P_1}{\partial a}\right)^2 \tag{A.60}$$

Note the fact that $E[u] = P_1$ has been used in (A.60). Since $P_1 = 1 - P_0$,

$$\left(\frac{\partial P_0}{\partial a}\right)^2 = \left(\frac{\partial P_1}{\partial a}\right)^2 = \frac{e^{-\frac{(\eta(\hat{a})-a)^2}{\sigma^2}}}{2\pi\sigma^2}, \tag{A.61}$$

where the relation

$$\frac{\partial Q[(\frac{\eta(\hat{a})-a}{\sigma})]}{\partial a} = \frac{e^{-\frac{(\eta(\hat{a})-a)^2}{2\sigma^2}}}{\sigma\sqrt{2\pi}} \tag{A.62}$$

has been used. From (A.59), (A.60), and (A.61), we get the desired result. □

A.14 Proof of Proposition 4.4

Let the probability of a sensor sending a bit value 1 be defined as P_1.

$$\begin{aligned} P_1 &= P(u=1|a, Byzantine)P(Byzantine) + P(u=1|a, Honest)P(Honest) \\ &= \alpha\left(p\left(1 - Q\left(\frac{\eta^B(\hat{a}) - a}{\sigma}\right)\right) + (1-p)Q\left(\frac{\eta^B(\hat{a}) - a}{\sigma}\right)\right) + \\ &\quad (1-\alpha)Q\left(\frac{\eta^H(\hat{a}) - a}{\sigma}\right), \end{aligned} \tag{A.63}$$

where $P(Byzantine) = \alpha$ is the probability that the sensor is a Byzantine, and similarly, $P(Honest) = 1 - \alpha$ is the probability that the sensor is honest. Here, p denotes the probability of flipping by the Byzantines. The data's contribution to the posterior Fisher information is given by

$$F = -E\left[\frac{\partial^2 \ln P(u|a)}{\partial^2 a}\right], \tag{A.64}$$

where $\ln p(u|a) = (1-u)\ln(1-P_1) + u \ln P_1$. Let $P_0 = 1 - P_1$. Then

$$\frac{\partial^2 \ln p(u|a)}{\partial^2 a} = -\frac{(1-u)}{P_0^2}\left(\frac{\partial P_0}{\partial a}\right)^2 + \frac{(1-u)}{P_0}\frac{\partial^2 P_0}{\partial^2 a} - \frac{u}{P_1^2}\left(\frac{\partial P_1}{\partial a}\right)^2 + \frac{u}{P_1}\frac{\partial^2 P_1}{\partial^2 a}$$

and

$$E\left[\frac{\partial^2 \ln p(u|a)}{\partial^2 a}\right] = -\frac{1}{P_0}\left(\frac{\partial P_0}{\partial a}\right)^2 - \frac{1}{P_1}\left(\frac{\partial P_1}{\partial a}\right)^2 \tag{A.65}$$

Note the fact that $E[u] = P_1$ has been used in (A.65). Since $P_1 = 1 - P_0$,

$$\left(\frac{\partial P_0}{\partial a}\right)^2 = \left(\frac{\partial P_1}{\partial a}\right)^2 = \frac{\left(-\alpha(2p-1)e^{-\frac{(\eta_B(\hat{a})-a)^2}{2\sigma^2}} + (1-\alpha)e^{-\frac{(\eta_H(\hat{a})-a)^2}{2\sigma^2}}\right)^2}{2\pi\sigma^2}, \tag{A.66}$$

where the relation

$$\frac{\partial Q[(\frac{\eta(\hat{a})-a)}{\sigma})]}{\partial a} = \frac{e^{-\frac{(\eta(\hat{a})-a)^2}{2\sigma^2}}}{\sigma\sqrt{2\pi}} \tag{A.67}$$

has been used. From (A.64), (A.65), and (A.66), we get the desired result.

Note that when $\alpha = 0$, the data's contribution to posterior Fisher information in (4.53) becomes

$$F[\eta(\hat{a}), a] = \frac{e^{-\frac{(\eta(\hat{a})-a)^2}{\sigma^2}}}{2\pi\sigma^2[Q(\frac{\eta(\hat{a})-a}{\sigma})][1 - Q(\frac{\eta(\hat{a})-a}{\sigma})]}, \quad \text{(A.68)}$$

which is the result of Proposition 4.2. □

A.15 Proof of Lemma 4.2

Let $d_H(\cdot, \cdot)$ be the Hamming distance between two vectors, for fixed $\theta \in R_j^k$,

$$\begin{aligned} P_e^k(\theta) &= P\left\{\text{detected region} \neq R_j^k|\theta\right\} \\ &\leq P\left\{d_H(\boldsymbol{u}^k, \boldsymbol{c}_{j+1}^k) \geq \min_{0\leq l\leq M-1, l\neq j} d_H(\boldsymbol{u}^k, \boldsymbol{c}_{l+1}^k)|\theta\right\} \\ &\leq \sum_{0\leq l\leq M-1, l\neq j} P\left\{d_H(\boldsymbol{u}^k, \boldsymbol{c}_{j+1}^k) \geq d_H(\boldsymbol{u}^k, \boldsymbol{c}_{l+1}^k)|\theta\right\} \\ &= \sum_{0\leq l\leq M-1, l\neq j} P\left\{\sum_{\{i\in[1,\cdots,N_k]: c_{(l+1)i}\neq c_{(j+1)i}\}} z_{i,j}^k \geq 0|\theta\right\}. \end{aligned} \quad \text{(A.69)}$$

Using the fact that $c_{(l+1)i}^k \neq c_{(j+1)i}^k$ for all $i \in S_j^k \cup S_l^k, l \neq j$, we can simplify the above equation. Also, observe that $\{z_{i,j}\}_{i=1}^{N_k}$ are independent across the sensors given θ. According to (2) in [2],

$$\begin{aligned} \lambda_m = \frac{1}{d_{m,k}} \sum_{i=1}^{N_k} (c_{(l+1)i}^k \oplus c_{(j+1)i}^k)(2q_{i,j}^k - 1) &= \frac{1}{d_{m,k}} \sum_{i\in S_j^k \cup S_l^k} (2q_{i,j}^k - 1) \quad \text{(A.70)} \\ &= \frac{1}{d_{m,k}} \left(\sum_{i\in S_j^k \cup S_l^k} 2q_{i,j}^k - \frac{2N_k}{M}\right) \end{aligned}$$

since $c_{(l+1)i}^k \neq c_{(j+1)i}^k$ for all $i \in S_j^k \cup S_l^k, l \neq j$. $\lambda_m < 0$ is then equivalent to condition (4.70). Therefore, using Lemma 4.1 and (A.70),

$$P\left\{\sum_{\{i\in[1,\cdots,N_k]: c_{(l+1)i}\neq c_{(j+1)i}\}} z_{i,j}^k \geq 0|\theta\right\} \leq \left(1 - \frac{\left(\sum_{i\in S_j^k \cup S_l^k}(2q_{i,j}^k - 1)\right)^2}{d_{m,k}^2}\right)^{d_{m,k}/2} \quad \text{(A.71)}$$

Substituting (A.71) in (A.69), we have (4.71). Note that condition (4.70) ($\lambda_m < 0$) implies $\lambda^k_{j,\max}(\theta) < 0$ by definition. Hence, (4.72) is a direct consequence from (4.71). □

A.16 Proof of Theorem 4.1

First, we prove that condition (4.70) is satisfied by the proposed scheme for all θ when $\sigma < \infty$. Hence, the inequality (4.72) can be applied to the proposed scheme. The probabilities $q^k_{i,j}$ given by (4.74) are

$$q^k_{i,j} = \begin{cases} 1 - Q\left(\frac{(\eta^k_i - a_i)}{\sigma}\right), & \text{for } i \in S^k_j \\ Q\left(\frac{(\eta^k_i - a_i)}{\sigma}\right), & \text{for } i \in S^k_l \end{cases}. \tag{A.72}$$

By Assumption 4.1, there exists a bijection function f from S^k_j to S^k_l. The sum $\sum_{i \in S^k_j \cup S^k_l} q^k_{i,j}$ of (4.70) can be evaluated by considering pairwise summations as follows. Let us consider one such pair ($i_j \in S^k_j$, $f(i_j) = i_l \in S^k_l$). Hence, their thresholds are $\eta^k_{i_j} = \eta^k_{i_l} = \eta$. Then, from (A.72),

$$q^k_{i_j,j} + q^k_{i_l,j} = 1 - Q\left(\frac{(\eta - a_{i_j})}{\sigma}\right) + Q\left(\frac{(\eta - a_{i_l})}{\sigma}\right) \tag{A.73}$$

$$= 1 - \left[Q\left(\frac{(\eta - a_{i_j})}{\sigma}\right) - Q\left(\frac{(\eta - a_{i_l})}{\sigma}\right)\right]. \tag{A.74}$$

Now observe that, by the assumption,

$$a_{i_j} = \frac{\sqrt{P_0}}{d_{i_j}} > \frac{\sqrt{P_0}}{d_{i_l}} = a_{i_l}$$

and, therefore, $Q\left(\frac{(\eta - a_{i_j})}{\sigma}\right) > Q\left(\frac{(\eta - a_{i_l})}{\sigma}\right)$ for all finite values of σ. From (A.74), the sum $q^k_{i_j,j} + q^k_{i_l,j}$ is strictly less than 1. Therefore, the sum $\sum_{i \in S^k_j \cup S^k_l} q^k_{i,j} < \frac{N_k}{M} = \frac{N}{M^{k+1}} = \frac{d_{m,k}}{2}$. Therefore, the condition in (4.70) is satisfied for the code matrix used in this scheme. Hence, $P^k_e(\theta)$ can always be bounded by (4.72).

By using (4.72), P^k_d can be bounded as follows:

$$P_d^k = 1 - \sum_{j=0}^{M-1} P\{\theta \in R_j^k\} P\left\{\text{detected region} \neq R_j^k | \theta \in R_j^k\right\}$$

$$= 1 - \frac{1}{M} \sum_{j=0}^{M-1} \int_\theta P\{\theta | \theta \in R_j^k\} P\left\{\text{detected region} \neq R_j^k | \theta, \theta \in R_j^k\right\} d\theta$$

$$= 1 - \frac{1}{M} \sum_{j=0}^{M-1} \int_{\theta \in R_j^k} P\{\theta | \theta \in R_j^k\} P_e^k(\theta)\, d\theta$$

$$\geq 1 - \frac{1}{M} \sum_{j=0}^{M-1} \int_{\theta \in R_j^k} P\{\theta | \theta \in R_j^k\} (M-1) \left(1 - \left(\lambda_{j,\max}^k(\theta)\right)^2\right)^{d_{m,k}/2} d\theta$$

$$\geq 1 - \frac{M-1}{M} \sum_{j=0}^{M-1} \left(1 - \left(\lambda_{j,\max}^k\right)^2\right)^{d_{m,k}/2} \int_{\theta \in R_j^k} P\{\theta | \theta \in R_j^k\}\, d\theta \tag{A.75}$$

$$\geq 1 - \frac{M-1}{M} \sum_{j=0}^{M-1} \left(1 - \left(\lambda_{\max}^k\right)^2\right)^{d_{m,k}/2} \tag{A.76}$$

$$= 1 - (M-1) \left(1 - \left(\lambda_{\max}^k\right)^2\right)^{d_{m,k}/2}. \tag{A.77}$$

Both (A.75) and (A.76) are true since $\lambda_{j,\max}^k < 0$ and $\lambda_{\max}^k < 0$. □

A.17 Proof of Theorem 4.4

First, we prove that when $\alpha < 0.5$, then

$$\sum_{i \in S_j^k \cup S_l^k} Z_i^{jl} E[\psi_i^k | \theta] \to \infty, \tag{A.78}$$

where $Z_i^{jl} = \frac{1}{2}((-1)^{c_{ji}^k} - (-1)^{c_{li}^k})$. Based on our code matrix design, Z_i^{jl} for $i \in S_j^k \cup S_l^k$ is given as

$$Z_i^{jl} = \begin{cases} -1, & \text{for } i \in S_j^k \\ +1, & \text{for } i \in S_l^k \end{cases}. \tag{A.79}$$

By using the pairwise summation approach discussed in Sect. 4.1.3.1, notice that, for every sensor $i_j \in S_j^k$ and its corresponding sensor $i_l \in S_l^k$, when $\theta \in R_j^k$,

$$Z_{i_j}^{jl} E[\psi_{i_j}^k | \theta] + Z_{i_l}^{jl} E[\psi_{i_l}^k | \theta] = E[(\psi_{i_l}^k - \psi_{i_j}^k) | \theta]. \tag{A.80}$$

Now, for a given sensor i,

$$E[\psi_i^k|\theta] = P(u_i^k = 0|\theta)E[\psi_i^k|\theta, u_i^k = 0] + P(u_i^k = 1|\theta)E[\psi_i^k|\theta, u_i^k = 1] \quad (A.81)$$

$$= (1 - P(u_i^k = 1|\theta))E[\psi_i^k|u_i^k = 0] + P(u_i^k = 1|\theta)E[\psi_i^k|u_i^k = 1] \quad (A.82)$$

$$= E[\psi_i^k|u_i^k = 0] + P(u_i^k = 1|\theta)\left[E[\psi_i^k|u_i^k = 1] - E[\psi_i^k|u_i^k = 0]\right], \quad (A.83)$$

where the facts that $P(u_i^k = 0|\theta) + P(u_i^k = 1|\theta) = 1$ and that the value of ψ_i^k depends only on u_i^k have been used.

Note that the channel statistics are the same for both the sensors. Therefore, $E[\psi_i^k|u_i^k = d]$ for $d = \{0, 1\}$ given by

$$E[\psi_i^k|u_i^k = d] = E\left[\ln \frac{P(v_i^k|u_i^k = 0)}{P(v_i^k|u_i^k = 1} \middle| u_i^k = d\right]$$

is the same for both the sensors. The pairwise sum $E[(\psi_{i_l}^k - \psi_{i_j}^k)|\theta]$ now simplifies to the following,

$$E[(\psi_{i_l}^k - \psi_{i_j}^k)|\theta]$$

$$= E[\psi_i^k|u_i^k = 0] + P(u_{i_l}^k = 1|\theta)\left[E[\psi_i^k|u_i^k = 1] - E[\psi_i^k|u_i^k = 0]\right]$$

$$- E[\psi_i^k|u_i^k = 0] - P(u_{i_j}^k = 1|\theta)\left[E[\psi_i^k|u_i^k = 1] - E[\psi_i^k|u_i^k = 0]\right] \quad (A.84)$$

$$= \left(P(u_{i_l}^k = 1|\theta) - P(u_{i_j}^k = 1|\theta)\right)\left[E[\psi_i^k|u_i^k = 1] - E[\psi_i^k|u_i^k = 0]\right]. \quad (A.85)$$

When $\theta \in R_j^k$,

$$P(u_{i_j}^k = 1|\theta) = \alpha + (1 - 2\alpha)Q\left(\frac{(\eta - a_{i_j})}{\sigma}\right) \quad (A.86)$$

$$P(u_{i_l}^k = 1|\theta) = \alpha + (1 - 2\alpha)Q\left(\frac{(\eta - a_{i_l})}{\sigma}\right) \quad (A.87)$$

and, therefore,

$$P(u_{i_l}^k = 1|\theta) - P(u_{i_j}^k = 1|\theta) = (1 - 2\alpha)\left(Q\left(\frac{(\eta - a_{i_l})}{\sigma}\right) - Q\left(\frac{(\eta - a_{i_j})}{\sigma}\right)\right). \quad (A.88)$$

Note that, since $\theta \in R_j^k$, $Q\left(\frac{(\eta - a_{i_l})}{\sigma}\right) < Q\left(\frac{(\eta - a_{i_j})}{\sigma}\right)$. Next, we prove that

$$E[\psi_i^k|u_i^k = 1] - E[\psi_i^k|u_i^k = 0] < 0 \quad (A.89)$$

for all finite noise variance of the fading channel (σ_f^2).

$$
\begin{aligned}
&E[\psi_i^k|u_i^k=1]-E[\psi_i^k|u_i^k=0]\\
&=E\left[\ln\frac{P(v_i^k|u_i^k=0)}{P(v_i^k|u_i^k=1}\middle|u_i^k=1\right]-E\left[\ln\frac{P(v_i^k|u_i^k=0)}{P(v_i^k|u_i^k=1}\middle|u_i^k=0\right]\\
&=\int_{-\infty}^{\infty}P(v_i^k|u_i^k=1)\ln\frac{P(v_i^k|u_i^k=0)}{P(v_i^k|u_i^k=1}dv_i^k-\int_{-\infty}^{\infty}P(v_i^k|u_i^k=0)\ln\frac{P(v_i^k|u_i^k=0)}{P(v_i^k|u_i^k=1}dv_i^k\\
&=-\left[D(P(v_i^k|u_i^k=1)||P(v_i^k|u_i^k=0)+D(P(v_i^k|u_i^k=0)||P(v_i^k|u_i^k=1)\right], \qquad (A.90)
\end{aligned}
$$

where $D(p||q)$ is the Kullback–Leibler distance between probability distributions p and q. Since $P(v_i^k|u_i^k=1)\neq P(v_i^k|u_i^k=0)$ for all finite σ_f^2, we have $D(P(v_i^k|u_i^k=1)||P(v_i^k|u_i^k=0)>0$ and $D(P(v_i^k|u_i^k=0)||P(v_i^k|u_i^k=1)>0$. This concludes that $E[\psi_i^k|u_i^k=1]-E[\psi_i^k|u_i^k=0]<0$. Hence, when $\alpha<1/2$, from (A.85), (A.88), and (A.89), $E[(\psi_{i_l}^k-\psi_{i_j}^k)|\theta]>0$ and the condition $\sum_{i\in S_j^k\cup S_l^k}Z_i^{jl}E[\psi_i^k|\theta]\to\infty$ is satisfied.

We now show that when the condition (A.78) is satisfied, the proposed scheme asymptotically attains perfect detection probability.

$$
\begin{aligned}
\lim_{N\to\infty}P_D&=\lim_{N\to\infty}\prod_{k=0}^{k^{stop}}P_d^k\\
&\geq\prod_{k=0}^{k^{stop}}\lim_{N\to\infty}\left[1-\sum_{j=0}^{M-1}P\left\{\theta\in R_j^k\right\}P\left\{\text{detected region}\neq R_j^k|\theta\in R_j^k\right\}\right]\\
&=\prod_{k=0}^{k^{stop}}\lim_{N\to\infty}\left[1-\frac{1}{M}\sum_{j=0}^{M-1}\int_\theta P\left\{\theta|\theta\in R_j^k\right\}P\left\{\text{detected region}\neq R_j^k|\theta,\theta\in R_j^k\right\}d\theta\right].
\end{aligned}
$$

Define

$$P_{e,j,\max}^k\triangleq\max_{\theta\in R_j^k}P_{e,j}^k(\theta) \qquad (A.91)$$

and

$$P_{e,\max}^k\triangleq\max_{0\leq j\leq M-1}P_{e,j,\max}^k. \qquad (A.92)$$

Then,

$$
\begin{aligned}
\lim_{N\to\infty}P_D&=\prod_{k=0}^{k^{stop}}\lim_{N\to\infty}\left[1-\frac{1}{M}\sum_{j=0}^{M-1}\int_\theta P\left\{\theta|\theta\in R_j^k\right\}P_{e,j}^k(\theta)d\theta\right]\\
&\geq\prod_{k=0}^{k^{stop}}\lim_{N\to\infty}\left[1-\frac{1}{M}\sum_{j=0}^{M-1}\int_{\theta\in R_j^k}P\left\{\theta|\theta\in R_j^k\right\}P_{e,j,\max}^k d\theta\right]
\end{aligned}
$$

$$= \prod_{k=0}^{k^{stop}} \lim_{N\to\infty} \left[1 - \frac{1}{M} \sum_{j=0}^{M-1} P_{e,j,\max}^{k} \int_{\theta \in R_j^k} P\left\{\theta | \theta \in R_j^k\right\} d\theta \right]$$

$$\geq \prod_{k=0}^{k^{stop}} \lim_{N\to\infty} \left[1 - \frac{P_{e,\max}^{k}}{M} \sum_{j=0}^{M-1} 1 \right]$$

$$= \prod_{k=0}^{k^{stop}} \left[1 - \lim_{N\to\infty} P_{e,\max}^{k} \right]. \tag{A.93}$$

Since $E\left[(\tilde{\psi}_i^k)^2|\theta\right]$ is bounded as shown by Lemma 4.4, Lindeberg condition [1] holds and $\frac{1}{\sigma_{\tilde{\psi}}(\theta)} \sum_{i \in S_j^k \cup S_l^k} Z_i^{jl} \tilde{\psi}_i^k$ tends to a standard Gaussian random variable by Lindeberg central limit theorem [1]. Therefore, from (4.95),

$$\lim_{N\to\infty} P_{e,j}^{k}(\theta) \leq \lim_{N\to\infty} \sum_{0 \leq l \leq M-1, l \neq j} P\left\{ \frac{1}{\sigma_{\tilde{\psi}}(\theta)} \sum_{i \in S_j^k \cup S_l^k} Z_i^{jl} \tilde{\psi}_i^k < -\frac{1}{\sigma_{\tilde{\psi}}(\theta)} \sum_{i \in S_j^k \cup S_l^k} Z_i^{jl} E[\psi_i^k|\theta] \middle| \theta \right\}$$

$$= \sum_{0 \leq l \leq M-1, l \neq j} \lim_{N\to\infty} Q\left(\frac{1}{\sigma_{\tilde{\psi}}(\theta)} \sum_{i \in S_j^k \cup S_l^k} Z_i^{jl} E[\psi_i^k|H_j^k] \right). \tag{A.94}$$

Since, for a fixed θ, $\sigma_{\tilde{\psi}}(\theta)$ will grow slower than $\sum_{i \in S_j^k \cup S_l^k} Z_i^{jl} E[\psi_i^k|\theta]$ when $\sum_{i \in S_j^k \cup S_l^k} Z_i^{jl} E[\psi_i^k|\theta] \to \infty$, $\lim_{N\to\infty} P_{e,j}^{k}(\theta) = 0$ for all θ. Hence, $\lim_{N\to\infty} P_{e,\max}^{k} = 0$, and from (A.93), $\lim_{N\to\infty} P_D = 1$ for all finite noise variance. □

References

1. Feller W (1966) An introduction to probability theory and its applications. Wiley, New York, NY
2. Yao C, Chen PN, Wang TY, Han YS, Varshney PK (2007) Performance analysis and code design for minimum Hamming distance fusion in wireless sensor networks. IEEE Trans Inf Theory 53(5):1716–1734. https://doi.org/10.1109/TIT.2007.894670

Printed in the United States
By Bookmasters